BACKYARD HOMESTEADING

MONA GREENY

Table of Contents

Backyard Homesteading
*Making Bread, Cheese, Drinkable Water
and Tea from Home*

Backyard Homesteading
Growing Vegetables, Fruits, and
Raising Livestock in an Urban House

Backyard Homesteading
Growing Flowers and Beekeeping in an Urban House

BACKYARD HOMESTEADING

Making Bread, Cheese,
Drinkable Water and Tea from Home

MONA GREENY

BACKYARD HOMESTEADING

Introduction

Everyone loves the idea of growing and producing food. It is not only cheaper but also healthier and more easily accessible. If you are on board with this idea of growing your food and eating healthy, home-cooked meals every day, then backyard homesteading is right up your alley.

If you have a knack for gardening, then you should definitely try building a backyard homestead. However, if you don't, then you will need a little practice and some patience to make it work.

Backyard homesteads are tailored to fit and for you to grow your food and other produce. It's a growing trend that is becoming more and more popular over time.

With this book, you can prepare your backyard homesteading system and reap the benefits of making your own products such as bread, cheese, drinking water, and tea.

Before we hop onto the process of building a backyard homestead, let's first take a look at the numerous benefits that come from building a backyard homestead. It should be enough to convince you.

What is Backyard Homesteading?

There are many ways to define a homestead. Historically, according to the Homestead Act of 1962, homesteads were pieces of land allotted to US citizens who agreed to settle in the West and live there for five years through farming. This definition evolved, and people perceived it as a productive concept that could be accomplished at home. Over time, homesteading aligned with the importance of living a self-sufficient lifestyle, which furthered the concept.

Homesteading is the activity of growing your food and produce like crops, vegetables, fruits, bread, honey, water, tea, etc. and using it for cooking your meals daily. Backyard homesteading is when you utilize your backyard to build a homestead and carry out these activities. On a higher level, homesteading experts even raise animals for meat and clothes. Produce like medicinal herbs, homemade medicinal products, plants and herbs for personal care, etc., are being included in homesteads. Basically, with homesteads, homeowners are trying to 'live off their land,' save money, and encourage a healthy lifestyle. One can call it an agrarian lifestyle.

Why You Should Consider Backyard Homesteading

With backyard homesteading, you are not only giving yourself a self-sufficient lifestyle but also entering a healthier, safer, and sustainable living domain.

Here are a few reasons that will compel you to prepare your backyard homestead:

1. Give and Take

You are giving something to the earth, and it is giving back to you - it's a symbiotic relationship. By building and maintaining a backyard homestead, you are contributing to the environment and enhancing the sustainability factor. Homesteading includes natural and sustainable procedures, such as composting and rainwater harvesting. By composting, you are making the soil richer and more nutritious, which in turn, is returned in the form of healthy and nutritious produce. By homesteading, you are reducing the carbon footprint and encouraging people to support sustainability.

Due to this, homesteaders are also more environment-conscious. They are closer to nature and are grateful to the earth that grows their food. This naturally instills a feeling of being environmentally conscious and taking accurate and sustainable measures in whatever they do. With the rising climate, excessive deforestation, and pollution, every person needs to take sustainable measures and natural incentives to protect their surroundings. The awareness and practice of building a backyard homestead can be inculcated by incorporating a self-sufficient lifestyle.

2. Promotes a Healthy Lifestyle

It goes without a doubt – growing your food in a backyard homestead provides healthy ingredients. Most commercially-sold crops and produce are infected with pesticides and fertilizers that are harmful to our bodies. These build toxins, which eventually result in bad health and diseases. This is why many people are switching to organic produce, which eliminates the use of artificial fertilizers and pesticides. With a backyard homestead, you can grow and consume your organic produce that is perfectly healthy and nutritious.

We all need some form of physical activity every day to stay healthy and avoid obesity. Since homesteading is a labor-intensive job, you are getting a lot of physical exercises daily, which further adds to the health factor.

Homesteading will not only keep you in shape and boost your physical health, but you will also begin to see changes in your mental health, too. It is noted that maximum homesteaders face less stress (especially the ones living in the countryside) compared to those who don't practice a self-sufficient lifestyle. The latter stay in urban areas with a lot of noise and pollution that adds to stress, which can be partly resolved by practicing backyard homesteading. It provides a sense of achievement and gratitude, which keeps your mental health at bay.

3. Builds a Self-Sufficient Lifestyle

Building a backyard homestead is the best way to create a self-sufficient lifestyle and stick to it. You can not only grow your food and use it, but you can also preserve it for the off-season. Moreover, you can also choose to grow food and crops that are useful for medicinal, personal, or skin care purposes.

It helps you connect with your surroundings and existing space in your vicinity. In addition, backyard homesteads give you the opportunity to explore more options than just growing vegetables and fruits. With this system, you can make bread, drinking water, cheese, and tea as well, making it multipurpose and multifunctional.

4. Saves a Lot of Money

Needless to say, by growing your food, you are saving a lot of money, too. The only investment you need to make is the tools required to set up a backyard homestead in the initial phase and the planting seeds. Even though it can be expensive in the beginning (given the hefty prices of equipment, tools for setup, and ingredients), you can save a lot of money in the long run. Additionally, by growing tons of food in your backyard, you can sell them to make more money.

5. Fresh Ingredients on the Go

It's a great idea to have a backyard homestead because you can get fresh ingredients at any time, all within your reach. With fresh ingredients and fresh food, you are blessing your body and keeping away diseases and health conditions. Commercially-bought produce has a lot of toxins and harmful fertilizers, which are difficult to get rid of. By growing your food, you are providing natural manure for your food to grow, which makes it fresh and healthy. Plus, if you live in the countryside, driving to and back from the grocery store can take a lot of time of effort. However, with food growing in your backyard, you no longer have to worry about commuting to the grocery store so often.

6. Makes for a Productive Hobby

Individuals who have been following a self-sufficient lifestyle through homesteading have turned hard-core, full-time homesteaders who try to live with zero waste and grow all of their produce. It is much easier for people living in open spaces and the countryside as urban areas have limited open spaces. Because of this, most urban homesteaders take it up as a small-scale hobby and to find something exciting to do at the end of their hectic day. It can be deemed as an escape from the 9 to 5 cycle. Irrespective of where you live and the area you possess, homesteading can enhance your lifestyle by multifold and offers a relaxing pastime. It also enhances your gardening skills.

7. Exemptions

If you live in the United States, you are offered homestead exemptions where you can sustain the value of your house and property and exempt yourself from paying some amount of taxes. This exemption continues even when a spouse dies as long as you keep living on the property. The advantage is called forced sales immunity, where you aren't forced to sell your property to cover your debt when you default on a loan. However, this doesn't make you immune to certain debts, such as mortgage foreclosure or defaulted property taxes. Before making a decision, consult a lawyer to understand all the clauses that apply to your state. This step by the government has encouraged many people to begin their own homestead and live a self-sufficient lifestyle, which also helps the environment.

8. More Self-Confidence and Bonding

Let's not forget the immense pride homesteaders have living on their land and growing their food. They also inspire others to live a self-sufficient lifestyle; this feeling of pride and gratitude makes them more self-confident and boosts their morale. Growing your food isn't easy, and by building a homestead, you can finally inherit the feeling of empowerment and self-control. You will also achieve a sense of progress and conduct. Moreover, it helps a person to learn from his mistakes and makes him more capable of overcoming challenges. However, for this feeling to sink in, make sure that you do not give up easily. You will fail once, you will fail twice, but with constant practice and patience, you will eventually make it, and this feeling of achievement will be priceless.

Also, with homesteading, you will finally get time to bond with your family. Everyone struggles to connect to their family due to different timelines and busy schedules. By taking up this project, you can meet and do things with your family or spouse, which helps you connect with them at a deeper level. If you have kids, include them in this practice as well. It's great to teach your kids the value of being environmentally conscious at a young age. Lastly, you can cook meals and try new recipes with your family with the produce you grow in your backyard homestead. It's a productive and motivating way to spend some quality time with your family and to chat about your daily whereabouts.

You see, there are several reasons why you should build a backyard homestead. If you are intrigued and cannot wait to build one, read on to garner the entire step-by-step process to begin today. From gathering the necessary tools and equipment to growing your own and raising, making and preserving your food, the following chapters will highlight every detail that you will need to build a successful backyard homestead. It might sound challenging and overwhelming in the beginning, but if you follow the steps with precision, you can achieve a successful backyard homesteading system in no time.

However, before you dive into it, think about it carefully, and ensure that you want to go through with it. The idea of building a backyard homestead surely seems alluring, but it's not as easy as it sounds. You need to put in a lot of time, effort, resources, and money (at least in the beginning) to start a successful backyard homestead. Moreover, most of the tasks involved in backyard homesteads are labor-intensive, which makes it more challenging. If you lack physical strength, homesteading is not for you, unless you have someone to do the laborious tasks for you. Homesteading with your partner is a wiser option as it divides responsibilities and makes management easier. For this, make sure that your spouse is 100 percent on board with the idea and tasks. Constant discussions, construction procedures, and tending to your homestead will become a regular part of your lives.

It is difficult, but not impossible. If you are intrigued and determined, don't wait for the perfect time; start today. Let's learn how to go about it from scratch - you will need the right facts and knowledge to implement this project into action, which the next chapters are all about.

Chapter 1

Gathering the Necessary
Tools and Equipment

Planning and preparing a backyard homestead is a lengthy and cumbersome process. There are so many steps involved and so much to take care of. However, if you take it one step at a time, you can easily achieve a backyard homestead that many dreams of. For this, we have broken the steps into chapters. The first step is gathering the necessary tools and equipment, which this chapter is all about.

It requires several tools and equipment.

Here is a list of tools that you will need to prepare your backyard homestead:

Hand Tools

1. Ax: If you plan to use a wood-burning stove in your backyard, you will need an ax to chop the wood. You can buy wood that is already split, but it will be too expensive and will increase your budget.

2. Shovel: Since you will be planting and growing herbs and vegetables in your backyard, you will need a shovel to dig. Use a shovel with a pointier edge to dig up harder parts of the ground.

3. Tool kit: A tool kit with all the necessary wrenches, screws, drill, wire strippers, retractable knives, etc., is one of the most important equipment in your checklist. If you don't have one, it is wiser to invest in a high-quality toolbox as it will come handy even in the future.

4. 12" combination square: This is often a forgotten tool that is actually very handy. A combination square is useful in leveling squares, finding angles, milling lumber, or taking quick measurements, etc.

5. Rake: A rake is used to clear the ground, raking up leaves, removing dirt, and leveling the ground. You can also do some digging using a rake. It is particularly useful during the fall when the leaves are many and all over the place.

Electrical and Construction Tools

1. Generator: If you want to go all-in and live in a tiny house or in an RV to celebrate a self-sufficient life, you will need a generator. While a 2400w generator works, a 3000w is even more powerful. However, consider the portability factor before you choose one.

2. Sawmill: If you are cutting and preparing your wood, you will need a sawmill system to make the process easier. Whether it is milling the center of the timber or the uneven sides of a wood piece, a good electric sawmill will do the job for you.

3. Volt or Ammeter: This simple device shows the strength, current, and power of the electric flow in a machine or system. You can also read the resistance to current flow in a machine using this simple device.

4. Wood or PVC: You will need wood to construct your garden beds and build your fence. Since it will be exposed to harsh exteriors, you need high-quality wood, such as cedar or PVC that can withstand heavy pressures, harsh weather, and moisture.

5. Solar panels: This is an optional yet highly recommended equipment that can save a lot of money in the long run. If your house allows the set-up of solar panels, you should definitely consider it.

Safety Gear

1. Safety goggles: This is necessary to protect your eyes while working with delicate wood pieces and heavy-duty electrical equipment. You can expect sparks and tiny wood chips to float during the process that could potentially hurt your eyes in the absence of safety goggles.

2. Hard hat: More like a protection helmet on construction sites, a hard hat is necessary to protect your head from accidents and heavy blows. Even if it is a small scale project, using a hard hat is highly recommended.

3. Earmuffs: Using electric drills and sawmills can produce a lot of noise that is unbearable at times and could damage your ears. Buy a pair of high-quality earmuffs that block heavy noise and will protect your ears.

Gardening

1. Seeds: After deciding the type of plants, vegetables, flowers, herbs, and fruits you want in your backyard homestead, you will need their seeds or saplings. Visit a nearby nursery to find better choices if you are confused.

2. Manure: Even though we suggest composting for organic and nutritious produce, you should still invest in some organic manure in the beginning until your composting system is ready. You definitely need manure if you are unwilling to make a composting system; it is useful in growing healthy and nutritious produce.

3. Watering can: You need a watering can to water your plants daily.

4. Soil: Depending on the types of plants and vegetation you want to grow; you will need soil that matches their individual needs. Certain plants need specific soil and texture to grow. For this, you need to plan your vegetation type carefully and buy soil accordingly.

Apart from these, you will need –

- A measuring tape

- Extension cord

- Hammer

- Heavy-duty staple gun

- Multi-purpose silicone cable ties

- 1000 lb Ratchet straps

- Storage bins

- Chisel sets

- Ladder

- Crowbar

- Palm sander

- Hose

- G-clamp

- Wheelbarrow

- Humidifier

- Air compressor

- Pliers

- Screwdrivers

Chapter 2

Studying and Preparing
your Backyard Homestead

This is an extremely crucial phase, as your homestead's success depends on the way you approach your backyard's analysis and setup. Even though there is no 'exact' formula to prepare a backyard homestead (since all backyards, climatic conditions, town regulations, and availability of local resources vary), you can make your personalized formula by checking a few criteria.

For this, you need to begin by evaluating the strengths and challenges of your property, followed by studying the climate, measuring dimensions, listing project ideas, preparing a budget, designing a layout, constructing it, and finally making it attractive. Let's take a look at these criteria one by one to make your workable backyard homestead formula.

1. Evaluating the Strengths and Challenges of your Property

Every property and backyard is different, which makes it a challenging job to come up with a common homestead design. Some people are blessed with huge yard areas while aspiring urban homesteaders need to design their homestead in a tiny area. To make the best out of your available space, you must first measure your property and evaluate the strengths and challenges of building your homestead. Even if you have a smaller space, you can make a workable homestead. All you need is some clever planning and creativity.

To evaluate your property's strengths and challenges, consider a few questions.

- Where is the property located?

- What type of property is it?

- Do you already own a piece of land, or are you looking for one?

- Are you planning to move to another place anytime soon?

- Is your property rental or permanent?

These are some basic questions that you can begin with. Even if you live in a rental property but do not plan to move soon, it shouldn't let you stop from finally building your homestead. You can still construct a couple of garden beds and plant trees and crops for fruits and veggies. Other ways to do homesteading work in rental properties include using containers to grow your food, build a composting system, and build a water purifier. However, make sure that your landlord approves it and get his permission (preferably in writing) before you begin. He might like the idea and even encourage you to add more systems. Even if it isn't your permanent home, it shouldn't hold you back to live a self-sufficient life.

If you live in your own permanent house, what are you even waiting for? As soon as you are done reading this book, grab your coat and get all the necessary equipment to build your homestead today.

After finding accurate answers to these questions, wander around on your plot to evaluate its physical strengths and weaknesses; this will include the quality of soil, functionality, and overall look and feel. Can you visualize a homestead on your plot? Does it look good? If yes, that's another tick in the box of a positive evaluation.

2. Study the Climate

Your region's climatic conditions will directly impact your backyard homestead. It will determine the type of produce that is suitable in that climate (considering all seasons), sun's exposure direction on your property, the amount of rain that falls in your region on an average rainy day and season, and the wind direction. Depending on the results, you can determine which crops and produce would be the most suitable for your area. The amount of sunlight and shade that will fall on your garden beds, the amount of water you can collect through the rain in your rainwater harvesting system, and the wind direction affecting delicate crops are some safe indications, to begin with. You will gain a better understanding after some experience.

Many homeowners and aspiring backyard homesteaders make the mistake of diving directly into the layout planning and construction of a homestead without sitting back and observing the climate. Even if it takes a few days or weeks to determine the accurate climatic conditions in your region, don't worry, as it will make your project successful in the long run.

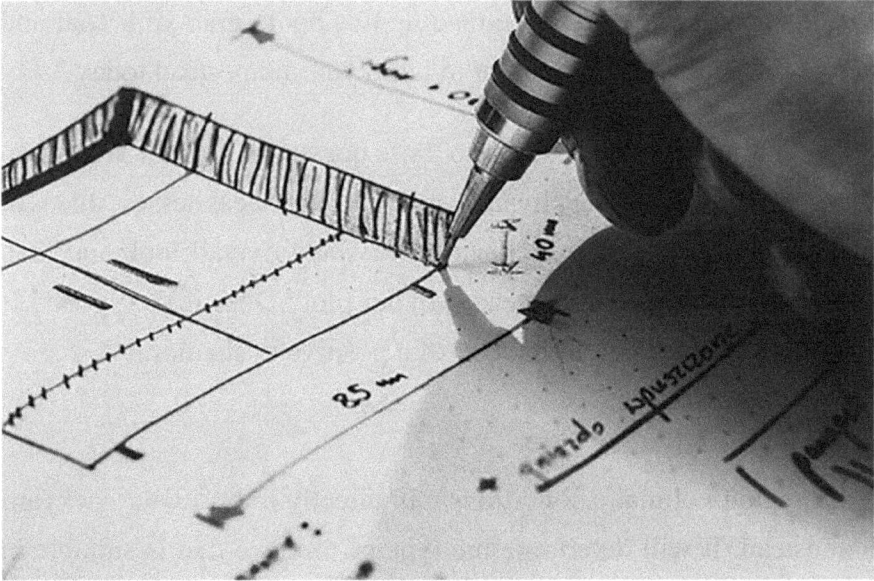

3. Measure Dimensions and Sizes

The size and dimensions of your property will matter greatly. Search for your house's plan or blueprint, or contact your architect to give you one if you have lost it. You can determine the exact dimensions of this plan. If not, you can use a measuring tape and get help from a friend or family member to determine the dimensions. Make sure that these are as precise as possible.

Measuring dimensions and sizes is also necessary to indicate the level of maintenance it will require. A medium or modestly-sized plot will surely have its limitations in the context of growing more produce, but it will be easier to maintain. Focus on longevity instead of quantity.

4. List Project Ideas

This section is about listing the ideas and objectives of your backyard homestead project. It mainly includes your requirements or produce that you want to achieve from your homestead.

Begin by making a list of objectives or produce that you wish to grow. Here are a few examples to get you started.

- An herb and vegetable garden with a rainwater harvesting system

- A backyard homestead that produces dairy - cheese, dairy, butter, etc.

- An orchard or a fruit tree (can also be combined with garden beds used for growing vegetables)

- A composting system with worms and a composting area

- An area dedicated to preparing seasonings, spices, and some condiments

- A medicinal plant garden that has only herbs and plants used for natural ailments, healthcare, and personal care

- An area that functions as a pollinator bed

- A bee-keeping area and a pantry dedicated to extracting honey

- A workshop to crochet, sew and work with natural dyes

- A composting system to make tea

- An area to package and sell your home-grown products and fresh ingredients

- A workshop area to teach and hold classes for other aspiring backyard homesteaders

These are some common additions to your backyard homestead that can be considered. Pick one or choose a combination as per your requirement and desire. Apart from this, you can also add your personal ideas. Just go wild (unless it's practically unachievable). These project ideas listed above are real, practical examples that hundreds of homesteaders are currently pursuing.

5. Prepare the Layout

Before you begin planning a layout, make a note of all rules and regulations that are applicable within your town, county, or state. When it comes to growing vegetables, rainwater harvesting, or raising bees for bee-keeping, every town has different regulations that you must abide by.

For the layout, you can, again, begin by asking a few questions.

- What are my requirements or produce goals from my backyard homestead? (It depends on the list of project ideas that we discussed above)

- Is it possible to fit all the desired produce within the plot? Is it big enough?

- If not, what produce can I eliminate, and what should I stick to?

- What are my urgent needs during the day?

- Will I pursue this as a hobby? Or do I want to go full-time?

It is advisable to pick only one to three ideas as it can get tiresome and overwhelming in the long run; it's practically not feasible, especially if you have a hectic schedule. Once you get the hang of it, you can then add more projects after one or two years of consistent homesteading. For multiple projects to work, you need to go all-in and offer full-time availability. You will also need additional help and a lot of resources for multiple projects. So, reflect on it and decide wisely. Set your priorities straight and take a relevant direction. Consider whether you want to take it up as a hobby or want to dive into it as a full-time profession (like selling your produce or hosting workshops).

At the same time, you should also consider the timelines. For example, until you have a well-organized pollinator system or a fruit garden, you shouldn't consider setting up a beehive as it goes hand-in-hand. Similarly, if you don't have an appropriate rainwater harvesting system that will take care of irrigation within the plot, you shouldn't plant herbs or crops that need a lot of water. All you will do is waste precious resources.

Also, consider the circumstances that will drive your backyard homesteading system. For instance, if your homestead lacks a well-planned irrigation system, the pipes can break at any point. For this, you will have to learn some plumbing skills or become an expert in this aspect. Similarly, if you don't know how to preserve large quantities of produce, all your hard work and grown food will go to waste. To avoid such circumstances, make them a priority, even if you have to learn specific skills to make the process smooth.

Once you have picked one or two manageable project ideas as mentioned above and acquired all the answers that will impact the layout of your backyard homestead, it is now time to design one.

If you live in a huge plot, you can divide the entire area into zones and work out the system accordingly. For example, since herbs and veggies are the most crucial ingredients for daily meal cooking, you can put it in zone 1, which is the nearest to your house. Similarly, the setup for drinking water should be close to your house, as you will need it quite frequently. Setups for making bread, cheese, and tea can be constructed away from your house but should still be accessible. This is known as permaculture design. Here, you notice the patterns and workability of your nearest ecosystem and design a homestead that works in harmony.

Other factors that will influence the planning stage include:

Access to water. You will need water in almost every stage of homesteading, from growing plants to making bread. Even though you have a rainwater harvesting system in the later stages, you will need a constant water supply in the initial stage. Determine your water supply (rainfall, access to nearby lake or pond, a well, etc.), and implement it in your project.

Neighborhood and community. Do you live in a neighborhood that shares a mutual wall or water supply line? If you share anything with your neighbor, you might want to ask their consent before you begin your homestead project.

Potential threats. Any kind of threat will ruin your homesteading project even before it begins. Threats like natural hazards (earthquakes or drought) or moles living under your soil can damage your homesteading system.

6. Making a Budget

Homesteading can be expensive, which is why it is crucial to prepare a budget before you begin construction.

Some plausible ways to save money include:

- Using second-hand materials and equipment or buying them at a discounted price. Since your equipment and construction materials will take up the maximum cost of the entire project, it is wiser to start saving from the initial stage. You can either borrow equipment from your close ones or rent them. Do not spend extra money on buying new tools as you only need to use it once (unless you decide to expand).

- Check out online stores such as Craigslist or Nextdoor, where you can find tools at a lower price. If need be, visit thrift stores in your area to get a good bargain.

- Up-cycle your possessions. This will not only save money but also provide a more sustainable alternative. For example, if you have used wood in your garage or barn, up-cycle it to construct garden beds or to make a fence.

Even though you can save money in various ways, it is always wiser to think twice before purchasing or renting an item. Certain things require investment due to quality variations. For example, a used, raised garden bed can be bought at a low price on an online store, but it won't last long. It might also be toxic or fail to pass the pressure test. To replace it, you will have to spend more money than you did initially, which is a significant loss. So, if need be, spend money to avoid unexpected costs as certain items need to last long. Think smartly and weigh quality over price.

Since wood will form the base of your garden beds and other structures, you need to ensure that it is durable, non-toxic, and withholds pressure. The best kind of wood (which lasts longer) is cedar and heart redwood. Even though these are a bit more expensive than their contemporaries, you won't have to spend another penny on wood for a few years, which makes it worth the investment.

To make a budget, list down all the tools and equipment you'd need to begin (take help from the previous chapter where we provided a list). Next, depending on what you want to include in your backyard homestead, make a list of equipment and ingredients to construct respective systems. For example, a vegetable growing patch will need materials to form a garden bed, soil, manure, and planting seeds. Calculate an estimate for each system; this will be your initial budget. Once your homestead is ready, prepare a list of ingredients and tools that you will require to maintain your homestead. Calculate the cost monthly.

Homesteading can be expensive, at least in the beginning. If you don't have a job and are planning to invest in the project with your savings, make sure that you have a proper return by selling your homestead produce. It is necessary to have an additional source of income to support your homestead and continue living a self-sufficient life.

7. Construction Phase

Even though we will talk about setting up specific produce systems for making bread, cheese, drinkable water, and tea (which are our main priorities) in the next chapters, you should at least have the basic knowledge of setting up a backyard homestead.

Fruits and Veggies

Apart from these commodities, you should definitely have a fruit and veggies area plus a tiny garden. A small fruit and vegetable garden sounds great in a residential context. It will provide fresh ingredients on the go and you can have a tranquil place to relax. Depending on what you grow, you could even create a cool, shady area to relax in too. Since ornamental trees and edible produce take some time to grow, you should start with this first. Focus on an organized irrigation system. Since most trees and veggies need a lot of water, a well-planned irrigation system should be your priority.

Composting System

Every homestead should have a composting system (if you have enough space). If you can, begin composting, you'll thank yourself later. Even if you are cramped for space, place a small composting system wherever you can. Starting a composting system will fall under the categories of living a self-sufficient and sustainable lifestyle, which is the main objective of homesteading. Place all your kitchen scraps and organic leftovers in a box to prepare your natural fertilizer. Use it to grow your healthy and entirely organic fruits and vegetables. You will be amazed by the quality of the produce at the end of each cycle.

Also known as black gold, natural compost will boost your soil's health and provide sufficient nutrients for healthy crops. Additionally, you have less trash to worry about. Food waste, such as used coffee grounds, vegetable and fruit peels, eggshells, etc., can be used in your composting system. Add some leaves, grass, or naturally-uprooted plants, and you are good to go.

Now, constructing your homestead might push you away from your comfort zone, but this experience will teach you a lot of new skills and might even turn you into a DIY enthusiast; it's really not that difficult. If you want, you can ask for help from your friends or hire an expert to build a basic skeleton. However, if you have all the tools, you can first try it yourself. It will not only save you a lot of money but will also give you a sense of achievement. Moreover, you will achieve that extra tinge of character by making your homestead from scratch. Lastly, you don't want to miss out on this amazing learning experience, right? When you build your homestead, you can also hold workshops or classes that teach aspiring homeowners to build their own backyard homestead.

8. Tips to Make it Aesthetic

Apart from constructing your homestead, you should also make sure that it looks and feels good.

Pay Attention to the Organization of the Layout

The most basic thing that makes a homestead aesthetic is the alignment of the components. When everything is organized and arranged, it automatically looks neat and feels good. For this, you should pay attention during the planning stage of your backyard homestead layout. For example, aligning the garden beds and providing adequate walking aisles in the middle seems organized. It is not only necessary to ensure proper air circulation and movement but also enhances the aesthetic appeal of the entire space.

Plant Flowers

This is one of the easiest ways to add an aesthetic appeal to any space. Even though your colorful garden beds and vegetables will make space attractive, adding a row of flowers will add a touch of flamboyance. When you are planning the layout, create an offset of minimum width in the borders of your property or backyard and dedicate that area to grow flowers. Your backyard homestead will not only look good but also feel and smell amazing. Another factor worth mentioning is that certain flowers also repel insects.

Play with Illumination

Considering that you will also tend to your backyard homestead at night, you will have to install a lighting and illumination system. Instead of using the regular fixture lights, you can take it up a notch and add hanging lights, which will enhance the entire space, provide visibility, and make the space utterly relaxing.

IMPORTANT NOTE: Since you are just starting with your first backyard homestead project, it is advisable to begin with a small and manageable project idea before jumping into multiple projects at once. Even if you want to choose only one idea, don't overwhelm yourself by overdoing things. For example, you might say to yourself, "I am choosing to grow only vegetables in the beginning, then why not plant six garden beds instead of three? How hard can it be?" This is the first mistake you are making. Stick to one or two project ideas and focus on smaller quantities in the beginning. What if it doesn't work out? What if you don't have enough time to tend to it? Since this is your experimental stage, you are bound to make multiple mistakes, which will result in a major loss of time, efforts, and resources. Instead, divert your attention to merely one or two projects in smaller quantities. Once these are successful, you can then move on to bigger projects and quantities.

While you are starting small and managing only one or two projects at a time, do not forget to leave room to expand in the future. Even though you will need only a small space, to begin with, you should comprehend the expansion possibilities once the initial phase is successful. This additional room will also give you space to amend your mistakes. You will keep on learning as you go, and mistakes are inevitable. You should learn from your mistakes and use space to improve your decisions. At times, things won't go as you anticipated, which is when this room for error and additional space will come to your rescue. Issues, like poor-grade soil and gophers climbing up and destroying your produce, are entirely out of your control. Additionally, what if you wake up one fine day and wish to change the layout or planning of your backyard homestead? So, prepare for the worst.

By making adjustments along the way, you are simplifying things. It is a better idea to take it one step at a time instead of going all-in at once.

To make it even simpler, pick a homestead partner or a work buddy who can support you along the way. The best person to go on this backyard homesteading project with is your spouse or roommate since they are always present. Any family member can make a great homestead partner as it needs someone's constant attention and care. Moreover, sharing responsibilities can take a significant load off your shoulders as your duties are equally divided. They help you stay motivated even when you don't feel like 'homesteading' on any particular day.

In the following chapters, we will take a look at some unconventional produce that can be made in your homestead backyard, which are bread, cheese, drinkable water, and tea. Growing vegetables and fruits is easy, and you must have a fairly good idea of how you can go about it. However, not many know that you can make other products with your backyard homestead, too, which is what the following chapters are all about.

Chapter 3

Make Your Own Bread

Once you have your backyard homestead ready, you are now ready to start making your produce. Let's start with bread. Bread is the staple food of every homestead because it is delicious, easy to make, and versatile.

A steaming, fresh loaf of bread is one of the easiest and most rewarding produce types that can be achieved from a backyard homestead. You no longer have to rush to the bakery in the morning and stand in a queue to get fresh bread. You can make it through a very simple process, which we will discuss in this chapter. Once you master the art of making bread, you can also sell them in fresh batches to earn money.

Basic Setup

To make homemade bread, you need a basic setup of a pantry or a kitchen platform. If you plan to make it in large batches, you need access to a bigger area. Make sure that it is near to the preservation or storage area for convenience and easy movement.

Tools and Equipment

Now, bread can be made in two ways - by kneading it with your hands or by using an electric dough mixer. Since we are talking about living off-grid, let's learn the most basic and best way of making fresh bread, which is by hand (because it saves energy).

To make fresh bread, you will need these tools:

- A large bowl

- ·Measuring cups and spoons

- A clean kitchen towel

- A big mixing spoon

- Oven and a deep baking dish

- Aluminum foil

- A large serving plate

- A cooling rack

- A sharp slicing knife

Ingredients

- 3 cups of flour (all-purpose or whole wheat)

- 1/2 tsp active dry yeast

- 1 tsp salt

- 1 1/2 cups of warm water

- Butter or oil (optional)

These ingredients will make one large loaf of bread that will serve around 10 to 12 people. You can double or triple the ingredients to make larger batches and sell them. The preparation time for one loaf of bread is around 2 hours, and pre-preparation is around 24 to 25 hours.

Process

1. Take a large mixing bowl and add in all the dry ingredients.

2. Add some water and mix it thoroughly. Add the water a little by little to avoid making lumps or making it too sticky.

3. Once all the ingredients are properly mixed, knead it with your hands until it forms a round and soft dough.

4. Press your fingers in the dough; if the dough bounces back and does not stick to your fingers, it is kneaded.

5. Cover the bowl with a clean kitchen towel and leave it for 8 to 24 hours. The longer you leave it to rise, the better results

you will achieve. For best results, leave it for around 24 hours. For this, you will need planning and action ahead of time.

6. After 24 hours, remove the cloth. You will notice that the dough has risen and is now twice the size with tiny bubbles formed on its surface. It will feel softer and spongier.

7. Spread some flour on a flat and clean surface (your kitchen platform is ideal for this) and place your dough on it. Put some flour on your hands and spread it evenly.

8. Lift the dough and make it into a ball. Knead it again to form an even texture. Once the dough is in the form of a ball, let it rest again for 30 minutes.

9. Meanwhile, preheat the oven at 450 F. Grease the baking dish with butter or oil. After 30 minutes, place the rested dough in the baking dish. Cover the top with aluminum foil and place it in the oven for 30 minutes.

10. Remove the foil carefully after 30 minutes and let it bake for another 10 to 15 minutes.

11. Remove it from the oven and place it on a cooling rack. Allow to cool for 10 to 20 minutes.

12. Slice the fresh and warm bread loaf with a sharp slicing knife and serve with butter or oil.

Pay maximum attention to the kneading process as it is the crucial stage that develops the gluten in your bread. To knead it properly, use the heels of your hands to push the dough with maximum force. Grab the dough, fold it, and repeat the process. Fold it once, knead it, and then pull it forward to make it softer and fluffier. Keep rotating the dough to exert equal pressure and to spread it evenly. Repeat it a few times. Continue the process for around 12 to 15 minutes before letting it rise. By kneading your dough properly, you will achieve fluffy and soft bread.

If you don't prefer kneading your dough with hands or lack practice, you can also use an electric dough mixer. A Kitchen Aid type of mixer is the most convenient in this case. All you need to do is add all ingredients in the mixer and let it do the kneading for you; this is particularly useful when you need to prepare large batches of bread to sell. Remove the dough from the mixer and knead it with your hands to provide that extra punch and force. With high powered mixers and appliances, such as Bosch, you can put in all ingredients, keep the setting at 1 or 2, and let it knead for 7 to 10 minutes. Remove the dough from the mixer, divide according to the number of loaves you want, and bake them. You don't have to worry about the dough rising with such high powered mixers as these allow additional air to enter while the kneading process is on. It reduces the resting time, and you get soft and fluffy bread at the end of each baking session.

Basic Questions

Bread-making is a craft that needs an ample amount of time and practice. It involves certain scientific steps that are beyond our apprehension. However, to master the skill, you should know the basics about these complexities. This section focuses on solving all basic and complex questions that might arise during bread-making.

1. What is the Difference Between Instant and Active Dry Yeast?

This is one of the most confusing questions for a beginner but is actually very simple. Yeast is an active ingredient in most bakery items that provide an airy texture to your bread and makes it softer. Yeast is a culture, which is dormant and comes to life when proofed or added to lukewarm water (active dry yeast). When added to the dough, it reacts with the ingredients and lets more air in, which results in softer and fluffier bread.

Instant yeast, as the name suggests, can be added directly to your recipe, whereas dry yeast needs to be activated beforehand. It is done by adding it to sugar and warm water. Active dry yeast feeds on the sugar to activate. Generally, instant yeast is preferred over active dry yeast as it is easier to use and takes no time. It is a finer version of active dry yeast in that it dissolves easily and activates in no time. You can also find it under the name quick-rise or rapid-rise yeast. These are types of instant yeast that are milled into even smaller particles to dissolve easily. These yeasts also have additives and enzymes to make the rising process quicker.

Rapid-rise yeast is perfect for quick baking projects that need to be accomplished within a day and still fetch perfect results. By using this type of yeast, you can skip the initial rising stage and jump to the step of shaping the dough.

On the other hand, active dry yeast has larger particles and needs to be activated in the initial stages of baking. It is usually sold in separate packets and is usually found in all grocery stores. You can also find this in small glass jars, which, when opened and used once, should be refrigerated to keep the yeast active.

2. What Kind of Yeast is Preferable for You?

It depends on the type of bread you are baking. It also depends on the amount of time you have to bake a certain amount in a batch. So, it ultimately depends on your situation and baking circumstances. Moreover, each type of yeast produces a different effect on varied baking items, which makes it even more difficult to come up with a definite answer. Try using all the kinds of yeast that you can get your hands on and experiment with different recipes. Compare the results and come up with one specific type of yeast that works the best for you. By doing this, you are also getting more familiar with the type that will produce better results. Get familiar with one kind of yeast that is versatile and readily available and use that one in all your baking projects. Unless any recipe suggests to you an alternative, stick to one yeast. Even if you can't find your specific yeast, don't fret; as long as you can find any brand of yeast, you are good to go.

The best part about yeasts is that they are interchangeable. If you are used to working with instant yeast but can't find one, you can easily switch to active dry yeast (provided that you follow the procedure). Also, if a recipe asks you to use some other kind, be flexible and adapt to the change as it will produce better results.

3. How Does the Bread Rise?

Once you decide to make your own bread, you should also know the scientific and methodical procedures that are involved in the process. One of the most important ones is the rising of the bread due to the addition of yeast. If you think about it, knowing the reason and science behind fermentation and the purpose of yeast is not that important. However, by learning the facts, you are getting closer to the fact and enabling yourself to get closer to the craft. This is why we will talk about the role of yeast in bread rising.

As you know, yeast is the main ingredient that is used for rising. It's a single-celled microorganism and exists in around 1500 types of species of fungi. Out of these, we use only one for baking purposes, which is Saccharomyces cerevisiae (it refers to the fungus type that eats sugar). This type, which is our baking yeast, is available in different forms, which mostly varies depending on its moisture content. We already know about instant and active dry yeast (which are commonly used for making bread), but there is a third kind, which is cake yeast. As you guessed, cake yeast is used for baking cakes and is compressed to remove excessive moisture; this allows us to form cake yeast into blocks. It works wonders in baking but can go bad within two weeks. It also enhances the flavor of the baked goods.

How yeast works – when you prepare the dough by kneading it, multiple air bubbles get trapped within the dough and spread throughout its internal composition. The activated yeast mingles with the sugar and starch present in the dough flour and turns them into carbon dioxide and alcohol. This release of carbon dioxide gas mingles with the already established air bubbles and inflates them, causing the bread to rise. The yeast's activity will multiply until there are enough carbohydrates and air for it to feed on. It will only reduce when it is exposed to the oven's heat - the reason why it is advised to keep the dough to rise for as long as possible.

Another interesting phenomenon that occurs during the rising is the growth of bacteria. It is similar to the phenomenon of the growth of lactic acid-producing bacteria in yogurt. This additional bacteria in the bread gives it an extra and delicious taste.

Now, many people presume that the rising takes place only after the dough is kneaded and left to rest. However, it takes place in two stages. The first is before you shape the loaf and knead the dough, and the second is when you leave it to rise. In the first stage, fermentation produces heat, which gets trapped in the center part of the dough. During this process, the yeast multiplies and produces alcohol and carbon dioxide gas, which eventually inflates the air bubbles. At this point, the activated yeast forms clusters. The process continues in the second stage during rising, where the yeast feeds on the food and air that is still present in the dough. It will continue until you pop the dough in the oven.

If you are flexible with the timing and can give more than 12 hours for the dough to rise, you will need less yeast. As the rising time increases, the less yeast you will need. However, make sure that you are still following the recipe. If a recipe asks you to add 2 tsp of yeast and recommends just 2 to 3 hours of rising, follow the instructions to fetch the best results. You can also tweak the measurements according to your schedule and preference. For example, if you are going out for a day, reduce the yeast amount by ½ or 1 tsp and let it rise for a day. Leave it in the refrigerator to complete the baking process when you return.

If you are leaving the dough to rise for more than 3 to 4 hours, you should place it in the refrigerator for better fermentation. By doing this, you are allowing the yeast to feed in a desirable environment without releasing its earthy flavor. In other words, fermentation in a cool spot triggers the release of more complex flavors that balances or cuts the yeast's earthy flavor.

4. What Can be Done if Your Quick Bread Recipe Doesn't Work?

You can turn it into muffins. Any quick bread batter can be turned into muffins using the same ingredients and measurements. Just make sure that you have the appropriate baking tray. In case you don't have one, use the bread baking tin to convert the batter into a cake. Add more milk or yogurt to liquefy its consistency, just enough to make a moist cake. Use any simple glazing, such as vanilla for topping or serve as it is. If the bread is too hard and dry, grind it in a processor to form breadcrumbs that can be used in other recipes.

5. Which Types of Bread Pans are the Best?

The best kinds are made of glass, ceramic, or cast iron, as these are durable and easy to use. These allow the heat to distribute evenly, which results in even baking. Also, it is easier to get the bread out of these types of pans. Some individuals, especially beginners, go with the non-stick kind as they seem easier for release. However, non-stick pans do not distribute heat evenly, which results in uneven baking.

6. What is the Difference Between Under Kneading and Over Kneading?

As you know by now, kneading is a crucial step in bread making. Over or under kneading can cause the bread to break or come out completely dry. The kneading should be 'just right.' The key to proper kneading is knowing when to stop. A little over or under kneading is allowed as it will still result in a fluffy and perfect loaf of bread. However, doing it too much or too little will ruin all your efforts, time, and resources.

Under kneading. Under kneading is when your dough breaks, tears, or is floppy. It doesn't have an even texture and will seem quite loose. To solve it, keep on kneading until you reach the perfect consistency - simple. To determine whether the dough is still under kneaded, try to shape it into a loaf. If it falls or fails to hold its shape, it is not yet done. If you have kneaded the dough and let it rise for a few hours, it will still not hold its shape if under kneaded. This is because the gluten that is forming the dough will not have enough elasticity to retain the shape. To solve under kneading after the dough has risen, shape it into a ball and let it rest for 15 to 20 minutes. Repeat the process until it holds the shape. You can also separate your dough into separate balls to make it quicker and easier. Once all separate pieces hold their shape, gather them together and to shape it into a loaf.

Make sure that you knead your dough properly as under kneading can make the resultant bread flat even after fixing it. The texture will be denser, and it will be difficult to slice the bread without tearing it. However, the taste will be retained. If it doesn't come out as desired, you can use this bread for other recipes such as French toast or bread pudding.

Over kneading. Over kneading usually happens when you are using a mixer to blend and knead the dough. It is quite impossible to over-knead your dough when kneading with your hands as it takes a lot of energy to do so. However, you still need to be careful as it is difficult to fix, especially when using an appliance. Pay attention to the electric mixer as it doesn't take much time to knead your dough. Stop it every 2 minutes to check the dough and touch it to determine the texture; this is particularly necessary when you are trying a new recipe or are new to baking. If your dough is over-kneaded, you will get a loaf of bread that is dense and dry from the inside and hard from outside.

If you are kneading the dough with your hands, stop when you feel that it is getting dense. Over-kneaded dough is dense and hard. Also, it is difficult to fold and flatten. The gluten formed during the kneading process will become tighter, which will make stretching it difficult. Due to this, you cannot fold the dough to knead it further, and it will eventually result in tearing of the dough. While it is difficult to fix over kneaded dough, you can still try by resting it for a longer time than initially intended. Even though the overworked gluten is irreversible, you can let the dough relax a little to enable it to retain the shape. Slicing the loaf will result in crumbly results. If the dough is extremely over kneaded, you can just use the bread for breadcrumbs.

Once you understand the difference and are able to distinguish between over and under kneading, you will achieve a beautiful loaf of bread after every baking session. Just observe and practice as much as you can.

To achieve perfect kneading, you can learn the windowpane test, which is an effective baker's technique to ensure that the dough is properly kneaded. It will prevent over kneading or under kneading of your bread. Before you take the windowpane test, go with your instinct to determine whether your dough has reached the proper consistency using the tips mentioned above.

Next, to check your dough using the windowpane test, cut out a small piece, about the size of a golf ball, from the kneaded dough. Place it between your thumb and two fingers. While the dough ball is still in place, stretch your fingers and thumb so that the dough ball stretches with it too. This stretched dough will take the form of a translucent membrane that resembles a windowpane. While the dough is being stretched between your fingers, check whether it is breaking or not. If it breaks or tears, your dough is not well-kneaded. Knead it again and check it using the windowpane test again. If it stretches and forms a translucent membrane, put it back in the main dough batch and shape it to form an even texture. Allow it to rise.

7. How Can You Make Your Yeast Last Long?

If left for a prolonged period, your yeast will expire. Imagine starting work on a big baking project only to find that your yeast has expired. Disappointing, right? The easiest way to sustain through such ordeals is to freeze your yeast. By freezing, your yeast will stay active even beyond its expiration date; this is because freezing puts the yeast's living cells into suspension instead of hurting them. To increase its effect, store your yeast in a dry and airtight container to keep out oxygen and humidity. A canning jar or a regular airtight glass jar will work. To use it, remove it from the freezer and place it in water, it will instantly come back to life.

To check whether the yeast is working perfectly, place it in lukewarm water with sugar. If you see bubbles forming in a few minutes, it is usable. This testing will prevent you from spoiling the other ingredients.

8. What are Some Useful Tips for Working with Yeast?

Using yeast can be tricky for beginners. It is one of the most crucial steps in bread making as it determines the texture and density of the baked bread. As we know, the two most common types of yeast are – instant dry yeast and active dry yeast. There are other commercial varieties of yeast that are used for purposes beyond baking, such as brewing beer, fermenting vinegar and soy sauce, and making wine. However, these come from select strains of the same yeast that are used for baking goods.

Ideally, you should test the yeast first before preparing your batch of bread. It will determine whether the yeast is working or not and save you from throwing away the entire batch. For this, using active dry yeast is recommended. Even though it involves an additional step at the beginning of the baking process, it will help determine whether the results will come out as anticipated. Also, the yeast hardly takes a few minutes and some effort to activate. As explained above, all you have to do is add it in lukewarm water and some sugar. If it starts bubbling in a few minutes, it will work perfectly well. This is the first step to ensure that your yeast will work, and your bread will come out fluffy and soft.

Secondly, you can also check the temperature of your workspace or kitchen to ensure the proper functioning of the yeast. Your yeast will work at its best when it is placed in a room with a temperature between 70 and 80 F. If you feel that the room is cooler or hotter than that, move the yeast mixture to an adequate space that has the right temperature for it to work. During winter, you can place the bowl of dough on a heater with some fluffy towels beneath it for effective rising. Keep checking on the bowl of dough during summer as it can rise quickly.

Another thing to consider is the number of ingredients you use in a bread recipe. For instance, if you are using sugar, eggs, butter, and milk all at once, it can increase the rising time. You don't have to worry about the yeast's expiration date. The additional ingredients just slow down the process of the yeast. When using such recipes, you should give it more time to rise.

With practice, you will learn how to handle yeast comfortably and extract the best results. Make sure that you stir, punch, and knead the bread dough adequately to release the alcohol and other toxic buildups that gives the bread a foul smell and bad taste. By kneading it properly, you even out the hot and cool temperature spots within the dough, allowing even rising and baking.

Important note: Before working with yeast, make sure that it is not dead as it could result in flat and tasteless bread. To determine whether the yeast is alive or not, open a packet, and smell it. It should have a typical earthy smell. To confirm its state, place it in lukewarm water and mix sugar in it. If it does not bubble up within 10 minutes, you shouldn't use it. Make sure that the temperature is accurate too. If it is below 70 F or above 130 F, the yeast will fail to activate. Do not let the temperature go above 138 F as it will kill the yeast.

Pay attention to the amount of salt, too, as it can affect its survival. If you add salt in the mixture in the initial stage, i.e., before the yeast has a chance to multiply, it will not show its true effect due to dehydration. Yeast needs water to thrive in an environment and enhance performance.

9. Does Shaping a Bread Loaf Need Special Attention?
Many people underestimate the importance of shaping bread. They either dump the bread in the baking tray without giving it shape or

do not put much effort into shaping their dough. Shaping the dough is crucial as it helps for even baking and gives an aesthetic appeal.

Here's the most basic and effective way to shape a loaf of sandwich bread:

- Shape the dough into a sphere with gentle movements. Then, flatten the sphere into a rectangle by gently pushing the dough with the heels of your hands. Use extra dough on the table surface or on your hands to avoid the dough from sticking to it. Don't use too much as it will lose its shape.

- Once you have formed a rough rectangle, imagine dividing it into three horizontal parts. Fold the bottom third part on itself, just like folding a letter.

- Next, fold the top part of the dough on the bottom to overlap it. Lastly, fold it again in half.

- The part where the top layer meets the bottom layer should be closed by pinching it with your fingers. Repeat the same with the sides.

- Once the loaf is shaped and ready, invert it in your baking sheet so the seams face down. Let it bake.

10. How Can You Determine Whether the Bread is Baked or Not?

While the easiest way to tell that the bread is baked is by looking at its color, it is usually baked when you see a nice golden brown color over its crust. However, at times, it can be deceiving. So, to determine whether the bread is baked or not, we will use three techniques for a full-proof check.

Visual. As mentioned, the color of any baked bread is a major indication of its baking stage. A deep, golden brown color with random dark spots is an ideal baked bread shade. At the same time, the crust is usually firm and dry. Do not remove the bread from the oven if the top still looks pale. If you are using a recipe, you will be told the ideal color of each type of bread. If not, keep practicing to learn to tell baked bread visually. With experience, you will be able to tell whether the bread is baked or not just by glancing at it. Until then, use the other two criteria as well.

Using a thermometer. Another way to determine the result is by checking the internal temperature of the bread you are baking. Stick a thermometer in the center of the bread loaf. If the temperature reads around 190 F, it is time to take it out of the oven as it is usually well-baked at this mark. Bread loaves with additional ingredients such as milk, eggs, or butter are usually baked at 200 F.

Checking the bottom. This is another simple way to determine the bread's cooked state. When you think it is baked, take the bread out of the oven and place it upside down. Use your thumb to strike the dough with a rapid thump (just like you'd hit the surface of a drum). If you hear a hollow sound, your bread is baked. You will need some practice with this technique as it can be a bit difficult to master. Strike the bottom of the bread at least three to five times after each baking session to determine whether it is baked or not.

Once you practice these techniques, you will gain more experience to determine the result intuitively. Practice more to learn faster. If you still have some doubt, leave it a bit longer in the oven - overcooking is better than undercooking. A dry crust will still be better than having an undercooked bread. However, instead of leaving it in the oven for an additional 5 minutes, do not make the mistake of leaving it for more than 15 to 20 minutes as it could burn the bottom.

If you feel that nothing works and you are still continually facing the issue of undercooked or overcooked bread, make sure that your oven is working properly. At times, ovens fail to work at the temperature that is shown, which could drastically affect the baking results. Place a thermometer inside to determine whether the oven is functioning at the right temperature or not. While a small change in temperature won't affect the results, you should get your oven checked if there is a difference of more than 50 degrees.

You will achieve a well-baked and beautiful loaf of bread after each baking session if you pay attention to these tips. With time and practice, you will figure out the particular technique that works well with you.

Types

The process that you learned above is basic bread making. Let's take it up a notch. With a backyard homestead, you can also make a variety of bread such as focaccia bread, no-knead whole wheat bread, sweet and savory bread, etc. In this section, we will take a look at three such bread types to kick-start your bread-making journey and turn you into a master. These are easy to make, easy to sell, and incredibly delicious. Let's take a look.

Focaccia Bread

Try this amazing Italian bread recipe that gives traditional bread a twist. You can also add pesto or enjoy plain focaccia bread on the side. Or, you can skip the pesto and use another filling of your choice. Focaccia bread is a typical Italian bakery item that is usually consumed with a savory meal. It's a great addition to your plate and goes well with the cheese that you will learn to prepare in the upcoming chapter.

To make this, you will need around 40 minutes of preparation time and 30 minutes of cooking time.

Equipment:

- A large mixing bowl

- Measuring cups and spoons

- A clean kitchen towel

- A big mixing spoon

- Oven and a shallow tin of 25 cm X 35 cm

- Parchment paper or baking sheet

- A large serving plate

- A cooling rack

- A sharp slicing knife

Ingredients:

The measurements provided here will serve around 10 to 12 people.

You will need –

- 20 ounces all-purpose or bread flour (plus some extra for dusting)

- 2 tsp salt

- 1 1/2 cups of warm water

- 0.25 ounces dried fast action or instant yeast

- 5 tbsps. of olive oil (plus some extra for greasing the baking pan)

- 1 tsp sea salt (flaky is more preferable)

- ½ tsp rosemary or ¼ bunch picked rosemary

Directions:

1. Take a large mixing bowl and add in all the flour. Add yeast to one side of the flour and add salt on the other side. This initial separation prevents the yeast from getting killed by the salt.

2. Add some water and mix it thoroughly. Add it little by little to avoid making lumps or making it too sticky. To make the process smoother, make a small well in the middle of the dry ingredients before adding the water.

3. Next, add 2 tbsps of oil and mix it thoroughly. Stop adding water when you feel that the dough is getting sticky. You need not use all the water.

4. Once all ingredients are properly mixed, knead it with your hands until it forms a round and soft dough. Press your fingers in the dough; if the dough bounces back and does not stick to your fingers, it is kneaded. Use the same kneading procedure mentioned above.

5. Cover the bowl with a clean kitchen towel and leave it for 8 to 24 hours. As we know, the longer you leave your dough to rest, the softer it will be. If you want the process to be faster, leave it for around 2 hours before you bake it.

6. Take the shallow tin and oil it. Spread some flour over it. Place the dough in the tin and spread it evenly. Cover it with the kitchen towel and let it rest for another 35 to 40 minutes to prove.

7. Meanwhile, preheat the oven to 450 F. Remove the towel after 35 to 40 minutes and press the dough with your fingers to form slight dimples on the surface. Push rosemary within these small holes so that they are flush with the surface.

8. In a separate small bowl, add 1 tbsp water, 1 ½ tbsp olive oil, and a pinch of flaky sea salt. Mix it evenly and spread it over the bread's surface.

9. Place it in the oven and let it bake for 20 minutes. It should turn golden by this time. Remove the bread from the oven and place it on a cooling rack. Drizzle 1 tbsp of olive oil on the loaf's surface while it is still hot. Allow it to cool until it has reached room temperature.

10. Slice the fresh and warm bread loaf into tiny squares with a sharp slicing knife and serve.

You can either serve this bread with pesto or plain. To add more flavor, add olives and onions in the dough before baking. It will provide a different texture and will make for an interesting flavor combination. Rosemary focaccia bread is the most common type of Italian focaccia bread. You can also serve it with salads.

Honey and Herb Bread

This is another interesting flavor combination that works best in bread. Honey and herb are like a match made in heaven, and with a bread recipe, these two completely stand out. The sweet and savory flavor combination works well with any meal and will definitely please your guests or customers. The best part is that you can harvest honey and herbs for this recipe from your backyard homestead. If you plan on selling your homemade bread, this one will surely be a hit. Let's learn how to make it.

Equipment:

- A large mixing bowl

- Measuring cups and spoons

- A clean kitchen towel

- A big mixing spoon

- Oven and a shallow tin of 25 cm X 35 cm

- Parchment paper or baking sheet

- A large serving plate

- A cooling rack

- A sharp slicing knife

Ingredients:

The measurements provided here will serve around 10 to 12 people.

You will need –

- 1 ½ cups of all-purpose or bread flour (plus some extra for dusting)

- 1 ½ cups of wheat flour

- 1/2 tsp salt

- 6 tbsp of warm water

- ½ cup regular water

- ¼ cup honey (plus some extra to drizzle on top)

- 2 pinches sugar

- 1 tsp dried fast-action or instant yeast or active dry yeast

- 1 ½ tbsp Fresh Thyme

- 4 tbsp Fresh Rosemary, Minced (you can also use mixed herbs)

- Butter (optional)

Directions:

1. If you are using active dry yeast: Take a small bowl and add yeast, sugar, and warm water. Stir it well and let it sit for around 10 minutes until it gets frothy.

2. Take a large mixing bowl and add the all-purpose flour, prepared yeast water, salt, and water. Mix it well. Add rosemary and thyme and remix it until well-combined.

3. Add some water and mix it thoroughly. Add it little by little to avoid making lumps or making it too sticky. To make the process smoother, make a small well in the middle of the dry ingredients before adding the water.

4. Once all ingredients are properly mixed, knead it with your hands until it forms a round and soft dough. Press your fingers in the dough; if the dough bounces back and does not stick to your fingers, it is kneaded. Use the same kneading procedure mentioned above. Knead it until the dough doesn't tear or break apart when pulled.

5. After the dough is kneaded properly, it is time to prove it. Place it in a big bowl and cover the top with a damp kitchen towel for 1 hour to prove. Make sure that the bowl is placed in a dry and warm place.

6. While the dough is resting, prepare the baking tray. Use a baking sheet or parchment paper to place over the tray and grease it with butter. After the dough has rested for 1 hour, knead it again with light hands and shape it into a circular dome.

7. Place it back in the dry and warm place and cover it with a damp cloth. Let it rest for another 30 minutes.

8. Meanwhile, preheat the oven to 350 F. Remove the towel after around 30 minutes and press the dough with your fingers to shape it into a circular dome. Then, use a sharp knife to draw a crisscross pattern on the top of the dough (three lines angled in one direction and two lines in the opposite direction). The reason behind drawing these lines is to make small troughs that can accommodate the honey without it spilling over.

9. Drizzle the honey in the gaps and place the dough with the baking tray in the oven.

10. Let it bake for around 35 to 40 minutes. It should turn golden by this time. Remove the bread from the oven and place it on a cooling rack. Drizzle some honey and apply some butter on the loaf's surface while it is still hot. Allow it to cool until it has reached room temperature.

11. Slice the fresh and warm bread loaf with a sharp slicing knife and serve.

You can serve this bread with cheese or just plain. It also tastes great with some cold lemonade.

Peach and Cream Bread

After two savory recipes, let's learn to make a sweet kind. This peach and cream bread recipe is a dessert delight and enjoyed by both adults and children. It's a simple flavor combination that tends to hit every kind of palate and makes for a decadent after meal dish. Moreover, this is a quick bread recipe, which, as the name suggests, can be baked within no time. Try this sweet and moist recipe today.

Equipment:

- A large mixing bowl

- Measuring cups and spoons

- A clean kitchen towel

- A big mixing spoon

- Oven and a shallow tin of 15 cm X 25 cm

- Parchment paper or baking sheet

- A large serving plate

- Absorbent tissues

- A cooling rack

- A sharp slicing knife

Ingredients:

The measurements provided here will serve around 6 to 8 people.

You will need –

- 2 cups of all-purpose or bread flour (plus some extra for dusting)

- 1 tsp baking soda

- 1 tsp salt

- 3 big peaches

- 3 oz cream cheese

- ½ cup vanilla Greek yogurt (or plain yogurt)

- 1 large egg

- 1 cup sugar

- 2 tsp vanilla extract

- For the vanilla glaze:

- ½ cup powdered sugar

- 1 tbsp milk

- 1 tbsp melted butter

- ¼ tsp vanilla extract

Directions:

1. Take a large mixing bowl and add in all-purpose flour, salt, and baking soda. Mix it well.

2. Cut the peaches into small pieces and dry them with a kitchen towel or tissue. Beat the egg in another small bowl.

3. Take a separate bigger bowl and add sugar, beaten egg, vanilla extract, and cream cheese. Mix it well.

4. When the mixture is well combined, add the yogurt and chopped peaches. Add this mixture in the flour bowl and remix it. Kneading is not required in this recipe as it is a quick bread recipe, and we want it to be moist.

5. Meanwhile, preheat the oven at 350 F. Prepare the baking tray. Use a baking sheet or parchment paper to place over the tray and grease it with butter.

6. Pour the dough mixture in the baking tray and spread it evenly. Let it bake for around 55 to 60 minutes.

7. Meanwhile, prepare the vanilla glaze for the topping. Take a clean bowl and add sugar, butter, milk, and vanilla extract. Whisk it well until it thickens.

8. Remove the bread from the oven and place it on a cooling rack. Allow it to cool until it has reached room temperature. Once it has cooled completely, drizzle the vanilla glaze over it.

9. Slice the fresh bread loaf with a sharp slicing knife and serve.

Ideally, this bread can be served during tea time or as dessert. If you are not a fan of peaches or want to add some variety to your bread collection, you can replace it with another fruit, such as apples, coconut, or bananas. To change the combination, experiment with additional ingredients like chocolate chips, cheese, cinnamon, etc.

Some flavor combinations that work well together include:

Apple + cinnamon

Banana + dark chocolate

Coconut + pineapple

Peanut + caramel

Strawberry + cream

Apricot + almonds

These are just some ideas to motivate you. You can also build a personal flavor combination by experimenting and adding it to your collection.

Preservation

Preserving bread is quite essential as it could develop mold or fungus when left unattended for a long time.

For Using at Home

The best way to use bread at home for a longer period is by freezing it. You can prepare large batches of bread on your free day and freeze them to use up to 3 months. Allow the baked bread to cool completely and wrap it in a plastic sheet. Wrap it again using an aluminum foil or place the wrapped bread in a freezer bag. Place it in the freezer. To use it, remove it from the freezer at least 12 hours ahead of time and leave it in the plastic wrapper to thaw. Within 10 to 12 hours, you will have fresh bread, ready to be used again. If the bread is still a bit hard due to freezing, sprinkle some clean water over the top until it soaks in. Bake it for 10 minutes at medium temperature. Prepare fresh glaze and drizzle on top or use it with melted butter.

For Selling

Needless to say, your bread should be fresh when placed for selling. Most bread expires after a period of 4 to 7 days, which is why you need to sell them as quickly as possible. The sooner you sell them, the better they will taste. If you think that the bread is about to go stale, wrap it and freeze it for your own use. Do not sell it. Your customers deserve fresh bread, so make fresh batches every 2 to 3 days; this will not only give you more customers but also increase your credibility.

Once you have your steaming and a fresh loaf of bread out of the oven, it is time to make some mouth-watering dishes with it. The simplest way to use your bread is to serve it as a side dish with pasta or any main dish. You can also spread butter or jam and have it for breakfast. But, why take the simple road when you can make it more interesting? You can convert your fresh loaf of bread into delectable recipes, such as a pesto and cheese focaccia sandwich, veggie burger, banana and chocolate chip bread pudding, etc.

Since you are building a homestead and planning on living a self-sufficient lifestyle, you might as well sell your produce, homemade bread, cheese, and tea. For this, you can invest in a heavy-duty mixer and a bread maker to make things easier and more concise for you. You can easily fetch the return of investment within a few months. By now you should have learned all the basics and tips of making your bread. Now all you need is some practice and dedication. If you continue practicing daily, you will master the craft in no time.

Chapter 4

Make your Own Cheese

Who doesn't love cheese? Imagine having a cheese workshop where you make and store your cheese daily. Sounds intriguing, doesn't it? With your backyard homestead, you no longer have to visit the cheese store and gauge at the varieties of cheeses. By making your cheese, you can enjoy different types of cheese depending on your preference and mood, and you will save a lot of money. Even though it needs some effort, it is a fun activity that can turn into your hobby or even profession. Not to forget the amount of money you can earn by making and selling your cheese. Whether it's a block of golden cheddar or decadent mozzarella balls, you can now have the cheese of your choice in the comfort of your backyard.

Also, you can enjoy a range of gourmet cheeses that are perfectly healthy and made from the best ingredients. A few commercial kinds of cheese are usually made of almost-spoiled or hormone-injected milk, which can be harmful to your health. However, by making your cheese, you have the complete freedom to choose your ingredients and make fresh cheese when you want.

Basic Setup

Depending on the types and quantity of cheese you want to make, you will need a wide workspace that can equip all the steps required to make cheese.

There are hundreds of varieties of cheeses that hail from different parts of the world. While some need a lot of processing and aging, other types are easier to produce and can be made at home.

For your basic setup, you will need a workspace where you can conduct the cheese-making process from scratch. You can either work in your kitchen or prepare a small pantry in your backyard that is accessible to the storage area. It should also have enough platform space to place several ingredients at once and conduct complicated procedures like separating the curds and pressing it.

Apart from the basic setup of boiling milk, separating curd, and pressing the cheese blocks, you will need a cellar and cheese cave to store and age your cheeses, which is thoroughly explained in one of the upcoming sections of this chapter.

Tools and Equipment

To make cheese in your pantry or kitchen workspace, you will need the following equipment, to begin with:

- A double boiler

- A stovetop or burner

- A food thermometer

- A cheesecloth or muslin cloth

These are some basic tools and equipment required to make simple cheese at home. However, if you plan to prepare various types of cheeses and in larger batches, you will also need the following equipment along with the ones mentioned above -

Measuring cups and spoons. Use glass or stainless steel measuring cups and spoons as these are non-corrosive and easier to sanitize. Do not use plastic as it can form scratches, which makes it more difficult to sanitize.

Thermometer. Since curd separates from milk at a certain temperature, it can be observed by using a thermometer more precisely. A good quality food thermometer will be useful in making bread as well. For better results, invest in a dairy thermometer. To check for its calibration, bring water to boil and check the temperature; it should read 212 F. If not, you should recalibrate it.

A cheese pot. A standard milk boiling pot will work perfectly fine. However, when buying one, make sure that it is made of a non-reactive material such as heat-safe glass or stainless steel. Do not use Teflon or aluminum as these could give rise to certain chemical reactions that will ruin the outcome. Check the size of the pot, too. If you plan to make larger batches, choose a pot that will accommodate large quantities of milk and allow you to stir and cut the curd comfortably.

Milk skimmer. Needless to say, a good quality milk skimmer with pores is efficient in stirring the milk and separating the formed curd. Use a stainless steel curd mixer for proper sanitization.

Cheese molds. Cheese molds are the easiest way to shape your cheese. These molds have holes that drain all the extra water from the curd. The two most basic cheese molds are made using food-grade plastic or stainless steel, as these are easier to handle and sanitize. Do not go for PVC molds as these could produce toxins in your cheese.

Colander. Use a big colander that is made of stainless steel or enamel and big enough to strain huge quantities of cheese curd.

Curd knife. Even though this is an optional tool, having a curd knife will definitely make the process easier. A curd knife is used in scraping the curd off the pot without damaging its texture or shape. You can use a kitchen knife too, but you should be extra careful due to its sharp edges. Choose a stainless steel knife that has a long blade to reach the bottom of any pot.

Drip tray. This is another optional yet useful tool that makes the process of cheese making more manageable and cleaner. A cheese drip tray drains out the excess whey from the cheese when it is pressed under a mold. You can place the drip tray beside the kitchen sink to let the whey drain directly into the sink.

Waxing pot and brush. Waxing is the process of coating an entire cheese block using wax to protect its surface during the aging process. For this, you will need a waxing pot and brush to coat the cheese surface with wax. While a double boiler is advised to heat the wax, you can opt for a heating pot to heat it with precautionary measures directly. A bristle brush is the best option to achieve an even spread. Make sure that the bristles are natural and not synthetic to prevent it from melting. Clean it immediately after waxing your cheese block and store it in a zip lock bag after every use. Do not use it for any other purpose.

Cheese wax. As mentioned, wax protects the cheese's surface and enhances the aging process due to the retention of moisture and balance. Cheese waxes are available in a variety of colors, the most obvious options being – yellow, red, black, and clear. Even though red wax is popularly used, you should opt for yellow wax as it provides visibility and allows you to see inside.

Cheese wraps. These are usually clear, breathable wrapping sheets that are used to store cheese and keep it fresh. You can choose between single, breathable wraps, or sheets made with two layers. These layers trap adequate amounts of moisture while allowing the exchange of gases for the best results.

Equipment to test pH and acid. Professionals generally use this equipment. However, if you desire to become a professional cheese maker in the comfort of your backyard, you can use this equipment to achieve the best results. The pH and acidity levels greatly vary during the cheese-making process, which can be checked and controlled with this equipment. It is particularly useful when making large batches of cheese after following a specific recipe.

Cheese press. This is another optional tool that can produce professional results if you are ready to make the investment. To make cheese, you need to hard press your curds to retain the shape and drain out excess whey; this is when a cheese press comes to your rescue. Make sure that your cheese press is easy to maintain, clean, and assemble. You will also find a gauge to understand the amount of pressure required to press a certain type of cheese. If you do not have an additional budget, you can simply use heavyweights, jugs, and books to press your cheese.

Cheese mats. Lastly, you will need cheese mats to drain excess whey and allow drying. These are usually made of thin food-grade plastic or reeds to allow breathability and passing of air. Cheese draining mats are necessary to block moisture accumulation, which will allow faster drying and aging.

These are some of the basic and professional tools used to make and age cheese. Apart from these, think about the cleaning and maintenance factors, too. For this, you will need a set of high grade sanitizing liquids to kill bacteria and provide a safe environment for the molds to grow. At the same time, you need the good bacteria to grow in the same environment. Cleaning and disinfecting the surfaces lowers the competition between bad and good bacteria, hence allowing the good bacteria to do its work and enhance the results. Make sure that you also clean and sanitize the equipment, storage shelves, and working surfaces to kill bad bacteria.

Ingredients

The most basic ingredients required to make simple cheeses at home are:

Milk (follow the next section to know the criteria in choosing the best quality milk): You can choose among raw milk, pasteurized milk, homogenized milk, ultra-pasteurized milk, cream line, dry milk powder, full cream, and non-dairy milk types.

Culture packs. Starter culture packs are the secret to great-tasting cheese. These are a group of friendly microbes that increase the acidity levels of milk and separate the curds. You can buy these in separate packs and are available in freezer dried form. If frozen, you can use these for around 2 to 3 years.

Rennet. Rennet is another essential ingredient used to make cheese. It is an enzyme that solidifies the proteins in the milk and coagulates it. Rennet is available in powder, tablet, and liquid form and can be easily found in supermarkets.

Additives. Apart from these basic ingredients, you will also need certain additives to enhance the flavor and texture of your cheese, such as cheese salt, herbs and spices, food coloring, acids, calcium chloride, and ash (activated carbon used for managing acid levels).

Process

The process that we will discuss in this section is one of the simplest ways to make cheese in your backyard homestead. It is divided into three parts – sanitization, preparing the curd, and storing the cheese to age. Let's take a look at these individually.

Sanitization

Before you begin with the actual step of boiling the milk and making cheese, it is necessary to have a safe environment that is free of harmful microbes and hygienic for adequate results. It is also necessary for the good bacteria to thrive and accelerate the process of milk ripening to separate curd.

Begin by sanitizing the tools and equipment that will be used in the cheese-making process. For this, you can fill your cheese pot with water and bring it to boil for 15 minutes. Place whatever equipment you can in the cheese pot and let it boil further; this will sterilize your tools and prevent health issues. Make sure that the surface you place these tools are sterilized too.

You can either use a sterilizing liquid or mix household bleach with distilled water to kill harmful microbes. Use disinfecting wipes or a clean cloth to wipe off the surfaces. Rinse it thoroughly as the bleach can also kill the good bacteria and rennet, which will result in bad cheese. You should not only clean and sterilize the surface and equipment at the beginning of the cheese-making process but also after it is done and before you store them.

Preparing the Cheese

Now, let's take a look at the main process of making cheese. This stage is the most crucial part of the entire process and can be further divided into numbered tasks to make it easier to understand and follow properly.

1. Preparing the Milk

The way you boil and prepare your milk is important because this stage is a primary factor in deciding the taste and texture of the final cheese. Choose a milk type that suits your needs (the following section will help you choose a milk type that is suitable for your needs) and add it in a huge pot to heat. If you are making a specific type of cheese, follow the temperature instructions given in the recipe. If not, let it heat on a slow flame. There are three different ways of doing this –

- **Water bath or double boiler method**. Fill your sink with warm water and place your pot filled with milk in the sink. Make sure that the water doesn't enter the pot. Check the temperature of the water using a thermometer and add boiling water if the temperature isn't adequate. It should read 10 F higher than the temperature of your milk. This method is effective in evenly heating the milk and keeps it from burning or scorching it.

- **Direct heating method**. The easiest way to heat milk is to place the pot directly over the stovetop. Pay attention to the temperature of the milk by using a thermometer as the chances of overheating are high. To prevent it from overheating or burning, use a pot with a thicker base.

- **Water jacketing**. This is more aligned to the double boiler method, where the pot of milk is placed in a water bath that is directly placed over the heat. Make sure that the larger pot is deep enough to hold enough water for heating the milk in the inner pot. It provides even heat distribution and prevents it from boiling or overheating.

The milk ideally must be heated at a temperature of 176 F to 200 F.

2. Making the Curd

This step involves adding various ingredients that will separate the milk curd from the whey. Now, the simplest way to make cheese at home is to use just two additional ingredients, which are salt and an acidic component (lemon juice or vinegar).

- When the milk is heating up, add a pinch of salt and turn off the flame when it reaches the desired temperature.

- Add some lemon juice or vinegar in the milk (around 1 tbsp per liter of milk) to separate the curd from the whey. You will notice the milk curdling or coagulating. Let it sit for at least 10 minutes for all the milk to separate into curd.

- Place a muslin cloth over an empty pot and strain the milk mixture to accumulate the curd on the cloth. Let it strain for an hour. Wrap the muslin cloth with the curd inside and squeeze it to remove excess water. Put it back to strain for another 30 minutes.

The remaining curd is your cheese, which tastes like cottage cheese. Shape the cheese using a cheese knife or some heavy pressure to form a cheese block.

For specific types of cheeses, you might need to add certain ingredients, such as food coloring, culture packs, rennet, and calcium chloride.

Starter culture pack. Depending on your cheese recipe, you will be instructed about the type of starter culture and rennet that must be added in the heated milk. You will also be given the time and temperature to follow. Before you mix the culture in the milk, let it rest on the surface for a minute or two to rehydrate it and prevent clumping; this will give an even mix and texture.

Adding rennet. Depending on your recipe, you will be further asked to add rennet. To mix it evenly in the heated milk, add the rennet in cool and distilled water and mix thoroughly. Add this rennet and water mixture in the milk. Follow the time, temperature, and resting period according to the recipe.

Adding mold powder. Some cheeses, such as Camembert, Brie, etc., need mold formation. For this, you will have to add a molding powder when you add the culture in the heated milk. The exact time and temperature will depend on the type of cheese you are making and its recipe. You can find several mold powders that vary according to their properties and quantities. Some of the commonly used mold powder types are red bacteria linens, white mold powder, and blue mold powder.

Adding citric acid. A few cheese recipes also call for the addition of citric acid and is typically used in fresh cheeses such as ricotta and mozzarella. It increases the acidity level in the milk without the help of starter cultures. For even distribution in the milk, add a small amount of citric acid in distilled and cool water and mix well. Pour this mixture in the heated milk and mix well again.

Adding lipase. Lipase is a taste-enhancing agent that is usually used in cheeses that have a mild taste. Lipase powders are extracted from animals. Two commonly used lipase powders are mild lipase powder (calf) and sharp lipase powder (lamb). It must be added to the heated milk before you add rennet (also if you are using a culture pack with rennet). For even distribution in the milk, add a small amount of lipase in distilled and cool water and mix well. Let it sit for around 20 minutes. Pour this mixture in the heated milk and mix well again.

Adding calcium chloride. Calcium chloride is used to form thicker and creamier curds. This is usually helpful when you use store-bought, pasteurized milk that lacks creaminess as in raw milk. This ingredient should be added before you add your culture and rennet to the milk. For even distribution in the milk, add a small amount of calcium chloride in distilled and cool water and mix well. Pour this mixture in the heated milk and mix well again. Do not add this ingredient when making stretchy cheeses, such as provolone or mozzarella, as it will ruin the texture and consistency.

Adding cheese coloring. This is entirely optional but can still be considered. As they say, you eat with your eyes first. If you want your cheese to look tempting and rich in flavor, you need a rich color. Colored cheeses like cheddar can use a few cheese coloring drops. If you want to use cheese coloring, add it to your milk, one drop at a time, and mix well. It must be added before you add your starter culture pack, rennet, and calcium chloride. Add one or two drops, mix well, and repeat until it has reached the desired shade. You should use around 20 to 50 drops for every gallon of milk. For even distribution in the milk, add a few drops of cheese coloring in distilled and cool water and mix well. Pour this mixture in the heated milk and mix well again.

Add one or more of these ingredients, depending on the type of cheese you are making. Let the milk reach your desired temperature and add the relevant ingredients accordingly. Let the milk coagulate and form curds, then let it rest for a while until all the whey is separated from the curd. For a better practical understanding, there are 4 types of cheese recipes in the upcoming section.

3. Accumulating and Cutting Curds

After letting the coagulated milk sit for a while, you will notice that the curd and whey have separated completely. Dry cheeses usually call for smaller curds, and moist cheeses usually need and form larger clusters of curds. The first step is to cut the curds, for which you will need a curd knife. Before you begin cutting the curds, you need to make sure that the formed curds have a 'clean break.' A clean break is when the curds easily separate when cut. To check for a clean break, insert your finger or the tip of your thermometer in the curds and check whether you get a clean break or not. If it separates, you are good to go. If not, let it rest for another 5 to 10 minutes until you achieve a clear separation.

Next, you need to cut the curds using the curd knife. For this, start by cutting vertical lines in the curds. Make sure that these are evenly spaced to achieve better texture results. Turn it to the other sides and cut perpendicular vertical lines that are equally spaced. You will get equal cubes of curd, just like a checkerboard. By doing this, you should notice a clear separation between the curds and a proper division between the whey and curds.

After cutting the curds, you might notice some 'whales' in your pot of curds. Whales are larger pieces of curds that are unequal in size. To fix the whales, let your curd rest for 5 to 10 minutes to make it firmer and easier to cut. Make sure that these are not too soft as it could disintegrate while cutting. Once you let it rest for a few minutes, stir it, and keep cutting the whales as they appear.

4. Cook the Curds

Some cheese recipes want you to cook the curd for complete separation of curds from the whey. It also balances the acidic levels and makes it firmer. The more the curd cooks, the smaller it becomes in size. Follow the recipe of the cheese you are making to achieve the temperature and time required to cook the curds. While some cheeses need some moisture to be left in the curds, others need it to be absolutely dry. The cooking time will depend on the moisture level and dryness required in the cheese. When cooking the curds, start with a slow flame and increase it by 2 F every 5 minutes; this will keep the curds from burning and overcooking. Make sure that you don't supply too much heat to the curds as it could form a skin and trap in moisture, which can result in uneven distribution of moisture in the curds.

Stir the curds continuously while they are cooking to prevent clumping and sticking to the bottom of the pot. Do not be too harsh in stirring as the curds are delicate in the initial stage. Breaking them with force will result in uneven texture. Depending on the type of cheese you are making, you might have to leave the curds in the pot to settle after cooking them. Let it cool down before straining it.

5. Draining the Cooked Curds

Some fresh cheeses, such as ricotta and cottage cheese, do not need the curds to be cooked as the moisture level and texture are adequate. All you have to do is let the curds settle for a while and for the whey to drain out. You can also squeeze the curds in a muslin cloth to remove excess whey. Let's take a look at it in detail.

The process of draining will depend on the type of cheese you are making.

For hard cheeses. Line a muslin cloth on your colander and place it in a sink. Use a ladle to transfer the curds from the pot to the colander. Let the whey drain in the sink for as long as the recipe suggests. Let it sit for a while until the curds are completely separated from the whey. Keep checking on the curds every once in a while. Due to uneven surfaces, some edges of the curds can be soupier. Scrape the sides of the cruds and gently blend it to achieve even draining.

For soft cheeses. Line a muslin cloth on your colander and place it in a sink. Use a ladle to transfer the curds from the pot to the colander. Tie the muslin cloth together to form a small packet and hang it on the sink tap to let it drain. Place a pot below to collect the whey or let it drain directly in the sink. You can also place the colander with a pot below it in the refrigerator. Even though it will take some time to drain the whey, you will achieve a great consistency in the end. Keep checking on the curds every once in a while. Due to uneven surfaces, some edges of the curds can be soupier. Scrape the sides of the cruds and gently blend it to achieve even draining.

For mold cheeses. Line a muslin cloth on your colander and place it in a sink. Use a ladle to transfer the curds from the pot to the colander. Let it drain for a while for the curds to completely separate from the whey. Mold cheese curds do not need much time to drain as they need moisture. Let it sit for a few minutes, and then you can directly transfer it to a cheese mold. The curds will compress themselves due to their own weight. Take a draining rack and place the cheese molds on it. Place a big pot underneath the draining rack or place it directly in the sink for the additional whey to drain out. You will notice that the compression of the curds will reduce their original size to half. Fill the cheese molds to the top and let it drain further. After a while (depending on the type of cheese you make and the recipe you follow), flip the cheese in the mold as per instructions. If your recipe calls for additional ingredients like dried herbs or ash, add it in the mold when it is half full. Once you add it, fill the rest of the mold with curd to the top.

Draining is a crucial stage as it determines the texture and consistency of the resultant cheese.

6. Pressing and Shaping the Cheese

To mold your curds, choose the shape and size of the molds as per your requirement. It will also depend on the quantity of cheese and the number of batches you are making. Line your mold with a muslin cloth and pour it in the curd. Fill it to the top as the curds will reduce by half upon compression. Place a follower on the mold to pack it. Next, you need to eliminate wrinkles by pulling out the edges of the muslin cloth.

Now, you need to press your cheese. If you have the luxury and plan on making several batches of cheese, it is wiser to invest in a cheese press as it is efficient and makes the process much easier. If not, you can use heavy objects such as books and pans to press the cheese. Cheese pressing removes excess whey from the curds and compresses it. The cheese block will reduce in size but will have an even texture and better consistency. Since the curds will form a tighter bond, you will also have a richer taste.

Depending on the type of cheese you are making, you will have to leave the cheese in the press between 10 minutes to 2 hours or more. Flip the cheese to repeat the process.

7. Salting the Cheese

Lastly, you will salt your cheese before preparing it for storage and aging. Salting the cheese will encourage rind formation and balance acid development in the cheese; this process is known as salt brine. Even though salt brining is not really necessary for homemade soft cheese, it can offer an extra level of protection for hard cheeses, especially if you plan on selling it commercially.

The process of salt brining will depend on the type of cheese you are making. For instance,

> **For hard cheeses**. Take a non-reactive pot and fill it with a brine solution to the top. To make a brine solution, mix 1-gallon water, 1 tbsp calcium chloride, 2 ¼ lbs salt, and 1 tsp white vinegar. Stir it well and add it to the pot. Place your pressed cheese block in the brine solution. Depending on the density of the cheese block, it might float in the pot. Since we need full immersion, you can add a pinch of salt on its surface. After a few minutes or hours of brining (depending on the type of hard cheese and recipe requirements), flip it to the other side for even brining.

For soft cheeses. To salt brine soft cheeses, there is no need to place it in a pot of brining solution. You can simply add some salt to the curd while it is draining; this will also force the removal of excess whey. Pay attention to the amount of salt you are adding as it could make the cheese extremely salty and ruin its taste.

For mold cheeses. When you place your cheese curds in the mold, you can apply salt to its exterior to balance the acidity levels and to drain off excess whey.

Aging and Storing

Once you have your soft or hard cheese ready, it is time to store and age it. Fresh cheeses can be easily stored in the refrigerator as these don't need aging. All you need is some cheese wrap and airtight containers to store these.

For hard cheeses and aging, follow the method provided in the next section that talks about preserving and aging cheese.

Basically, after you press and salt brine your cheese molds, let them air dry for a while. Remove them from the mold or muslin cloth and place them on a cheese drying mat. Use a cheesecloth or a muslin cloth to cover its surface when placed for drying. After it is left to air dry for some time, flip it on the other side and cover it with the muslin cloth again.

Before you store the cheese blocks for aging, consider waxing them for additional protection (follow the next section for the entire process). Even though waxing is the easiest option to retain moisture in the cheese, you can also consider bandaging certain cheeses like cheddar. To bandage your cheese block, cut two circular pieces of muslin cloth (washed and sanitized) that are slightly bigger than the top of your cheese blocks. Soak these cloth pieces in warm water and place them on the top and bottom of the cheese block. Place it back in the mold and repress with the cloth on. Let it press for about an hour or two. Once it is thoroughly pressed, cut the muslin cloth in the dimensions of the circumference of the cheese block. Soak it in hot water and place it on the circumference of the cheese. Spread it evenly so that no wrinkles or creases are formed. Place in the mold and press it again - this time, leave it in the cheese press overnight.

Lastly, you can apply olive oil or any other food-grade oil to your hard cheeses for extra protection. This is useful for protecting the cheese from exterior contaminants. Before you oil your cheese, make sure that it is thoroughly air-dried and cleaned. Remove any contaminants or mold that has grown on the cheese surface. Let it air dry once again. Once that is done, apply olive oil on the surface of the cheese with a bristle brush. You can also mix herbs, cocoa powder, or other flavoring agents to enhance the flavor. It is now ready to be stored and aged.

Basic Questions

Cheese-making is a lengthy process that can get complex in some instances. You will face certain why's and how's midway. To help you become a master cheesemaker, this section will answer all the basic and complex questions to clear all doubts.

1. How Can One Determine the Quality of Milk to Make Cheese?

As mentioned, the inexpensive store-bought cheese is often made of poor quality or hormonally injected milk, which can be bad for health. If you are a cheese lover and plan to make and sell your cheese from the comfort of your backyard, you will need high-quality materials. It will not only safeguard your family's health, but it will also help you gain credibility as a cheese producer and seller. The main ingredient that is needed to make cheese and needs to be high in quality is milk. But how can one determine the quality of milk, assuming that you bought it from your nearest supermarket?

Needless to say, the quality of milk affects the quality of cheese produced. It is further driven by specific criteria, such as – the way the milk is processed, when it is produced, its type, etc. It also depends on the type of animal from which the milk is extracted, such as cow, buffalo, goat, or sheep, etc. This is because the composition and nutritional quality of the resultant cheese will differ vastly. In fact, the quality of the milk used will also differ according to the breed of one type of animal.

Let's take each criterion that affects milk quality and understand them briefly.

- Time: The time taken to milk the animal, from drying them off to freshening.

- Season: The season during which the animal is milked also affects the milk quality due to the difference in milk solids that are produced.

- Exercise, diet, and health of the animal: Animals that lack exercise will produce poor quality milk compared to animals that exercise. The diet (feed and pastures) that is fed to the animal also matters.

- Location: Even though the type and breed of the animal are similar, there will still be a difference in milk quality due to different locations.

- Handling and care: Animals bred on farms on a small scale versus those bred in large commercial sheds vary greatly in the amount and type of milk produced. For example, smaller groups of animals on farms are personally tended to, which provides more care and attention to each animal.

- Processing or treatment: Different processes or treatments affect the quality too. There is a huge difference between raw and pasteurized milk.

- Milking: The way the animals are milked – through hands or automated equipment, also makes a difference.

- Storage: If chilled and appropriately stored, the produced milk will retain its taste and texture. Moreover, it also depends on the way the milk is stored, individually or in large storage tanks.

- Transportation: Milk brought from farms is recently milked and fresher than milk transported from big milk production setups.

After looking at these criteria, you can determine the quality of milk that is suitable for making your cheese batches. You should pay specific attention to these if you plan to make big batches and sell them since the quality of your cheese will be significantly affected.

If certain criteria are difficult to decipher, you can also get help from these parameters to begin working:

- Taste of the milk: Try different types of good-quality milk and choose one with the best taste and flavor.

- Processing: Do not buy milk that is pasteurized at a temperature above 168 F as it is not suitable to make cheese. The milk should be pasteurized at or below 162 F and should be held for at least 16 to 20 seconds at this temperature range.

- Expiration: Buy milk that is fresher and far from the best before date.

- Cost: Assuming that you will be making cheese for a longer-term, you should go for a milk type that is low in cost and affordable as it could affect your budget in the long run.

Lastly, experiment with a few milk types. If you are confused and unable to choose a type, this practical approach will determine a solid winner. Make small batches of cheese using the same technique and time but with different milk. The one that has the best taste and texture is the clear winner. If you are a beginner, you should opt for good-quality, pasteurized milk as it is free of microbes and for optimum cheese making. Raw milk has higher fat content, but can be risky to use due to the presence of unwanted microbes and might cause health risks. Even if you want to go for raw milk due to better curd formation and excellent texture, make sure that you get it directly from a nearby farm and use it within 2 to 3 days. After consuming raw milk, if any irritation or discomfort occurs, consult your doctor.

2. What are the Criteria to Select Good Cheese Wraps?

Most cheese making beginners often forget the importance of selecting a good cheese wrap as it affects the storage and freshness of the cheese for a prolonged period. While a cheese wrap should protect its contents from a harsh environment, it should also let fresh air in and moisture out for adequate ripening. An ideal cheese wrap is made of a single-layered breathable material (which is usually cellophane) that provides a controlled ambiance for the cheese to age and stay fresh at the same time. Since this type of cheese wrap is transparent, you can see the aging process and observe the changes on the surface.

The best kind of cheese wraps for soft and mold-ripened cheese are clear and made of minimum pores to lessen evaporation. However, the minimal amount of pores present on this cheese wrap allow gas (which is built during the aging process) to escape. For washed rind cheese, you can opt for clear wraps that are clear yet more porous. This is to keep moisture away from the cheese as it can ruin the aging process.

Another type of cheese wrap is a two-ply wrap, which provides more protection. If you are selling your cheese on a professional platform, this wrap should be your go-to. It not only keeps moisture away but is also capable of managing rough handling in retail buying. The inner layer of this cheese wrap absorbs moisture from the cheese surface, and the outer layer retains moisture but does not let it escape; this phenomenon occurs in the initial stage of ripening. As the cheese further ripens and reaches its later stage, it uses the moisture locked in by the second layer as required. Use a high-quality two-ply cheese wrap for proper moisture balance. If the inner layer absorbs excessive moisture, it could ruin the cheese by sticking to its surface.

This two-ply cheese wrap also varies according to the type of cheese you are storing. For example, mold-ripened cheeses like Brie, Camembert, and goat cheese can be stored in two-ply cheese wraps that are made of parchment paper (inner layer) and a micro-perforated polypropylene paper (outer layer). The inner layer is coated with paraffin and joined to the outer layer, which keeps moisture and mold away from the surface of the cheese. You can find another kind of two-ply cheese wrap that is useful for washed rind cheese. The sulphurized grease-proof sheet, which is the thin inner layer, keeps moisture away from the cheese surface, and the micro-perforated polypropylene paper, which is the outer layer, keeps gas flow intact. These two layers are sealed together tightly and offer transparency and visibility. In addition, these wraps specialize in preventing crystal formation, which is a common issue in grainy rinds.

For wet or moist cheese, the ideal wrap is a plain waxed paper sheet. It reduces moisture from the contents; it is wrapped around and offers the perfect taste and texture balance.

Why and How Should You Wax Cheese?

Waxing cheese provides a layer of protection for the cheese to keep its moisture level intact and to prevent mold growth. You can buy cheese wax, which is usually a microcrystalline wax type and safe on cheeses. It is crack resistant and prevents the cheese from forming bumps due to mishandling, and it also makes handling and maintenance easier. Some cheesemakers use paraffin and beeswax to wax cheese. However, these can crack easily and allow mold formation during aging.

To wax your cheese, remove all the mold from its surface and clean it thoroughly using a vinegar wash or a brine wash. The best way to wax cheese is on a low temperature as the mold spores are alive and will form mold even after the cheese is waxed. Take a bowl or a pot and place it in the wax to heat. Use a double boiler to prevent the wax from taking direct heat. Once the wax is fully melted, take a natural bristle brush to wax the cheese surface. Try to cover the cheese surface as fast as you can before the wax solidifies, but make sure that you do it with precision. Coat it with an even layer of wax, but do not overcoat it. Take one surface at a time. Wax it and let it solidify. Reheat the wax and apply it to another surface of the cheese block. Repeat until you have covered the entire cheese block with wax.

Take precautionary measures when heating and melting wax as the vapor produced could ignite a fire and result in life-threatening circumstances. Even though a double boiler heating method does not kill mold spores, it is more preferable due to the safety factor. At times, it is possible to find mold on the surface of the cheese even after it is waxed. This is due to improper cleaning of the surface with vinegar or improper wax coating that could have caused a hole on the surface.

4. How do Starter Cultures Work?

As mentioned, these are a group of friendly bacteria that separate the curd from the milk by fermenting the lactose content present. This phenomenon produces lactic acid, which changes the internal composition of the milk that is used. Depending on the fermentation level and grade, the mineral and moisture content changes accordingly; this, in turn, determines the taste and texture of the resultant cheese produced. You can either follow a recipe and add the starter culture pack as per the instructions or use your instincts after practicing it a few times. For accurate results, note the temperature of the milk when left to boil and add the culture pack accordingly. Culture packs show effect only at a certain temperature, which is why you should check it time and again. This will allow a proper environment for acid production and development, which will enhance the process of curd formation.

5. What are the Different Types of Starter Culture Packs and Which One is More Suitable?

In general, there are two kinds of starter cultures - mesophilic and thermophilic. These vary according to their difference in temperature tolerance. While mesophilic cultures need low heat to activate, thermophilic cultures operate at a higher temperature. If heated more, mesophilic cultures die as these cannot tolerate high temperatures. These two cultures are further divided into various types of cultures, which are used individually or as a combination of 2 to 4 to make cheese.

You need to choose a culture type based on the type of cheese you are making and the temperature at which you need to boil the milk. The best way to determine it is by following the instructions mentioned in your recipe. You can either use a direct set culture (used directly in the milk) or a re-culturable culture (needs you to create a mother culture in the beginning). The easiest and most convenient way to use a starter culture is by choosing a direct culture set as it requires less effort and involves less technicality. Some direct culture sets also have rennet in it to ease the process.

To use a re-culturable culture, heat the milk to reach 72 F for mesophilic culture or 110 F for thermophilic culture. As soon as it reaches your desired temperature, add the culture pack to it. Seal the container and let it sit for 6 to 8 hours. This setting provides a suitable environment for the microbes to multiply and form a mother culture. Open the seal, and you will find a gel-like consistency, which is your mother's culture. To store it, divide it into an ice tray, and freeze the cubes. Once frozen, store them in airtight freezer bags in the freezer and use them as required.

6. How Does Rennet Work and What are its Different Types?

As mentioned, rennet solidifies milk proteins and coagulates it to separate the curd from the whey. Even if you don't add rennet, it will solidify if left out for a couple of hours due to acid production. However, it will also create an off flavor at the same time. To retain its sweet taste, rennet is added to solidify the milk before excessive acid production, which also retains its texture.

There are many types and forms of rennet – animal, vegetable, liquid, tablet, powdered, and junket. Animal rennet, which is extracted from the fourth stomach lining of calves, is used more often. This is due to the presence of pepsin and rennin enzymes. However, with the rising concerns of animal cruelty, people are switching to vegetable-based rennet, which is derived from several sources, including Fig, Mucur Miehei, Nettle, Mallow, Knapweed, Yarrow, and Teasel. Common rennet forms include liquid, tablets, and powder. Liquid rennet is the best choice for beginners, as it is easy to measure and use. You merely need to measure, pour in the milk, and stir. It can be stored in the refrigerator and can last from 4 to 12 months (depending on the type of rennet). The sooner you use it, the better results you will achieve. Even though it doesn't go bad quickly, you should use it as soon as possible as it could lose its potency. If the rennet is stored for a prolonged period, you can double the amount to achieve effective results.

Powdered rennet needs a temperature range between 38 and 45 F for storage and lasts around one year. If you live in a tropical location with hotter days, you should choose powdered rennet over liquid rennet as it is easier to measure and not prone to constant fluctuations in temperature. If your use is limited and not that frequent, you can go for tablet rennet as it is long-lasting and can be used up to five years if frozen or up to one year if stored in the refrigerator. Make sure that you crush and dissolve the rennet tablets in water before adding it to the milk to distribute it evenly. Junket is another form of tablet rennet that is much weaker than regular rennet and is ideal for making soft cheeses.

Types

After learning the process of making simple cheese, let's, now, dig into various types of cheeses that can be made in your homestead set up easily, such as mozzarella, ricotta, and mascarpone.

Here's how you can make these cheeses at home:

Mozzarella

Most people stick to store-bought mozzarella due to the intimidating feeling of making it at home and failing to achieve perfect results. However, if you set your mind to it and follow this step by step guide, you can achieve creamy and stretchy mozzarella sticks in your backyard every day within 30 minutes.

Ingredients

- ¼ cup water (chlorine-free)

- gallon milk (you can use 1 percent, 2 percent, and even skim milk)

- ½ rennet tablet

- 2 tsp citric acid

These ingredients will make cheese that will serve around 6 to 8 people.

Equipment

- Measuring cups and spoons

- Boiling pot

- Ladle

- Mortar and pestle

- A large pot

- Dairy thermometer

- A sieve or muslin cloth

- Mixing spoons

- Stove or Induction

Directions

1. Clean and sterilize your apparatus and working station before you begin the main process.

2. Use the mortar and pestle to crush the rennet tablets into a fine powder. Add it to a small bowl and add water. Stir it until it dissolves completely.

3. Add milk in a pot for boiling and add the citric acid to it. Stir it well. Heat it until it reaches 88 F. You will notice the milk beginning to curdle.

4. Once the temperature has reached 88 F, add the crushed rennet powder and keep stirring the milk solution until the milk temperature reaches 105 F. Take the pot off the flame, and you will notice the formation of huge chunks of curd. The whey, which is then separated liquid in the reacted milk, will appear in the form of a clear and green liquid.

5. Place a muslin cloth or a cheese strainer on a large empty bowl and scoop out the curd using a ladle. Squeeze the curd

using your hands or a spoon to remove excess water. Let it strain for an hour.

6. Transfer the curd into a big glass bowl and microwave it for 1 minute. Once done, remove it and press the curd with a spoon with full force. Do not use your hands as the curd will be too hot.

7. This pressing phenomenon will remove extra whey that is still trapped in the curd. Repeat until the curd cools down. Place it again in the microwave and let it cook for 35 minutes. Repeat the pressing process.

8. You need three rounds of microwaving and pressing. Once all the extra whey is drained out, you will be left with just a ball of cheese. Knead it like bread dough until it loses a bit of its softness.

9. Place it again in the microwave for 10 seconds to remove the hardness. Remove it and knead it again. Repeat until the cheese forms a shiny and soft outer texture. While you are kneading the cheese, add a pinch of salt to it.

10. Now, this step is key to long, soft, and stretchy mozzarella. You need to pull the cheese like a rope and fold it. Do it repeatedly until the cheese comes to a soft, silky, and stretchy consistency. If the cheese tears in the middle of the stretching process, pop it back in the microwave and cook it for 10 seconds on high.

11. When you think that you have achieved a soft and stretchy consistency, shape it into a ball and serve as you please. You

can also add mixed herbs or chili flakes during the kneading process to give it an extra boost of flavor.

You can now prepare fresh Caprese salad or serve homemade fresh mozzarella cheese to your guests. With some practice, you can master the skill and add mozzarella to your commercial selling range of cheeses.

Mascarpone

Mascarpone is a soft and fresh cheese that is made using cream. It is used as the main ingredient in tiramisu and as a side for salads and meals. This cheese is a rich and creamy delicacy that can also be eaten as a spread.

Ingredients

- 1 quart pasteurized heavy cream

- ¼ tsp tartaric acid

These ingredients will make cheese that will serve around 6 to 8 people.

Equipment

- Measuring cups and spoons

- Boiling pot

- Ladle

- Mortar and pestle

- A large pot

- Dairy thermometer

- A sieve or muslin cloth

- Mixing spoons

- Stove or Induction

Directions

1. Clean and sterilize your apparatus and working station before you begin the main process.

2. Prepare a double boiler by fitting a saucepan over a pot to heat the cream. Heat it until it reaches a temperature of 180 F.

3. Once it reaches the desired temperature, remove it from the heat and stir it for 2 minutes. Add the tartaric acid and let the cream curdle.

4. Meanwhile, place a muslin cloth or a cheese strainer on a large empty bowl and scoop out the curd using a ladle. Let it sit for around 12 hours in a cool spot. You can also place it in the lower level of your fridge until it cools down completely.

5. Use four squares (of approximately 9 inches) and place the prepared mascarpone in the center of these squares. Fold each side and overlap to form tiny packages of mascarpone cheese. Refrigerate for 12 hours or more and serve.

Ricotta

Ricotta is another easy cheese to make at home. It tastes great with bread and in salads. It hardly takes a few minutes to prepare and can be deemed as a house staple.

Ingredients

- ½ gallon milk (you can use 1 percent, 2 percent, and even skim milk)

- 2 cups buttermilk

These ingredients will make cheese that will serve around 4 to 6 people.

Equipment

- Measuring cups and spoons

- Boiling pot

- Colander

- Rubber bands

- Ladle

- A large pot

- Dairy thermometer

- A sieve or muslin cloth

- Mixing spoons

- Stove or Induction

Directions

1. Clean and sterilize your apparatus and working station before you begin the main process.

2. Add milk and buttermilk in a boiling pot and stir it well. Do not let the mixture boil. While it is heating up for the first five minutes, stir it continuously to prevent overheating or burning the milk mixture.

3. Stir it until it reaches a temperature of 100 F. You will notice the milk beginning to thicken and curdle.

4. Take the pot off the flame when the temperature reaches 175 F, and you will notice the formation of huge chunks of curd. The whey, which is the separated liquid in the reacted milk, will appear in the form of a clear liquid. Let it sit for 5 to 10 minutes.

5. Place a muslin cloth or a cheese strainer on a large empty bowl and scoop out the curd using a ladle. Squeeze the curd using your hands or a spoon to remove excess water. Let it strain for an hour.

6. Once the whey drains out, grab the ends of the muslin cloth and tie them together with a rubber band to form a small pouch. Tie it on the handle of the ladle and let it hang over a pot to collect the drained whey. Let it drain for another 30 minutes.

7. Your ricotta recipe is ready to be served. Refrigerate it and use it within 3 to 4 days until it is fresh.

Cheddar

Apart from these soft and fresh cheeses, let's take a look at how to make a hard cheese at home to give you a kick-start. Cheddar is one of the tastiest cheeses that you will enjoy making and eating. Even though it's a long process, the result will make it worth the effort.

Ingredients

- ¼ cup water (chlorine-free)

- 2 gallons full cream milk

- 1 packet direct set mesophilic culture pack

- ½ tsp liquid rennet or ¼ tsp double strength liquid rennet

- 1/8 tsp calcium chloride

- 2 tbsp sea salt

These ingredients will make cheese that will serve around 6 to 8 people.

Equipment

- Measuring cups and spoons
- Boiling pot
- Long curd knife
- Cheese wax
- Cheese draining mat
- Ladle
- A large pot
- Dairy thermometer
- A sieve or muslin cloth
- Mixing spoons
- Cheese press
- Stove or Induction

Directions

1. Clean and sterilize your apparatus and working station before you begin the main process.

2. Pour milk in a pot and heat it until it reaches a temperature of 85 F. Add the calcium chloride at this point (you can dissolve it in some water and add it to the milk for even distribution). Add the starter culture once the milk reaches 85 F again, and the calcium chloride has dissolved properly.

3. Stir it properly, cover it with a lid, and let it ferment for around 60 minutes. Stir it again and add the rennet (you can dissolve it in some water and add it to the milk for even distribution). Stir it properly and remove the pot from the flame.

4. Let it sit for an hour. Keep checking the pot to notice the curd being separated from the whey. It should be distinctive. Once the curds and whey are entirely separated, use the curd knife to cut the curds into equal cubes of ¼ inches.

5. Let it sit for another 5 minutes. It is time to cook your curds. Make sure that you cook the curds on a slow flame as it could burn or disintegrate, which will ruin the cheese's composition and texture. Put it on a low flame and continuously stir until it reaches a temperature of 100 F; this step will take around 30 minutes.

6. Take it off the heat and allow it cool for around 20 minutes. Pour the curds into a colander and drain excess whey in the sink. Place it back in the cheese pot. Let it sit for 15 minutes until the additional whey drains out. Remove the curds from the pot and place them on a cutting board.

7. Cut it into slices and place these back into the pot. Cover with a lid. We will now form a water jacket or a double boiler by filling the sink with warm water that reads 102 F. Place the pot with the curds inside the sink so that the curds reach a temperature of 100 F. Turn the sides of these cheeses every 15 minutes. Continue this process for around 2 hours. Pay

attention to this step as it will determine the flavor of your resultant cheese.

8. By the end of this process, you will achieve firm and shiny slices of curds. Cut these into equal cubes measuring ½ inch cubes. Transfer these cubes back into the pot and place it in the sink with warm water that reads 102 F.

9. Leave it for around 10 minutes, stir gently, and leave it again for another 10 minutes. Repeat two more times. Remove the pot from the warm water and place it in a dry spot. Add some salt and stir thoroughly.

10. It is time to press the cheese. Prepare the cheese press by lining a cheesecloth over it. Transfer the curds from the pot to the cheese press. Cover all the surfaces of the curd using the cheesecloth. Press the curd at 10 pounds of pressure for around 15 minutes.

11. Once it is done, remove it from the cheese press, flip it on the other side, and place it back under the press. Make sure that you unwrap and rewrap the cheese with the cheesecloth with each flip. Use a fresh cheesecloth the second time. Apply a pressure of 40 pounds this time and let it press for around 12 hours. Repeat the entire flipping and pressing step once again. However, apply a pressure of 50 pounds this time and leave it for 24 hours.

12. Once the cheese is fully pressed, remove it from the press and place it in a cool spot to air dry for at least 2 to 3 days. Keep checking the cheese every once in a while and flip it at regular intervals.

13. Lastly, wax the cheese to provide extra protection and leave it in a cellar to age. The temperature of the cheese cave or cellar should be between 55 F and 60 F. The cheese will take around 60 days or more to age.

Preservation

Among other things, cheese needs special care and attention for preservation. Some cheeses are preserved to be aged as they smell and taste better after a certain period. Whether you are producing your cheese for consumption at home or producing it in large quantities to sell, you should prepare an apt setting to preserve and age your cheese to suit all purposes.

For Using at Home

Homemade cheeses expire quickly, especially if they are placed in a warm environment. For this, you should refrigerate your cheeses to increase shelf life. Depending on the type of cheese you are making, pack it in a wrapper and label them with the production date. Buy some cheese boxes to store your cheese batches safely; this will help you keep track of its freshness. You can also stick an expiration date or a 'use by' label. Use these cheeses while they are fresh and retain their best quality.

Before you pop your batches of cheese in the refrigerator, make sure that you clean your refrigerator compartment using a disinfectant, vinegar, or wine as they can go bad due to contamination. You can also put a glass of water in your cheese box to retain humidity. While placing different types of cheeses in your refrigerator, ensure that they are at a suitable distance to avoid cross-contamination. Check the cheeses every day and apply olive oil or coconut oil if they are getting dry.

For Selling

Making a cheese cave in your backyard is a great way to preserve and age your cheese. Cheeses have been stored and aged in literal caves for centuries due to the perfect humidity and temperature inside. Today, cheesemakers create tall sheds that resemble a cave's environment for their cheese to age with perfection. The good news is, you can make your own cheese cave in your backyard to store your cheese for aging.

Consider these three conditions to make your cheese cave:

- A consistent internal temperature of 45 F to 58 F

- Moisture level of 80 to 98 percent (depending on the types of cheese stored)

- Some fresh air to prevent unwanted product formation

Here are a few ways to make your cheese cave:

1. Use an Old Refrigerator

This is the easiest and most convenient way to store and age your cheese in large batches. Modify your old refrigerator or buy a new one to only store cheese. Even though refrigerators are regulated at a lower temperature (10 to 15 degrees below the cave temperature), you can use airtight containers to prevent cheese from drying and also to retain moisture. While choosing an airtight container, make sure that it is much bigger than the cheese block and provides 60 percent empty space for air. Crumple a wet paper towel and place it in one corner of the container to control humidity. You can also use a pan filled with water and a cover placed on top. Use a hygrometer to check the moisture level within the fridge because moisture levels will vary greatly as the season's change. These can be controlled using a damp towel or by spraying water.

2. Building Your Cheese Cellar

If you want to go all-in, you can build a cheese cellar in your backyard or within your pantry. A 15' X 25' cellar with a 6' X 15' room is enough to store your batches of cheese. Place an insulated wall with a vapor barrier around the 6' X 15' room to provide optimum temperature and humidity. To prevent sunlight, place an insulated door (made using urethane placed between plywood panels) in the northern direction. The internal temperature should stay at 52 to 54 F, and the humidity at around 85 to 90 percent in the 6' X 15' room (which is also your actual cheese cave). However, if you are located in a spot that gets too hot during summer or too cold during winter, you might have to place a heater in the bigger cellar room during winter or an air conditioner during summer to regulate the temperature. You can use the larger cellar space to dry your cheese before placing them in the cave to age.

Construct some shelves in the cellar and cave to place your cheese. The best wood combination to construct these shelves is pine and ash, as it is easy to work with and can be cleaned easily. Make sure that these are sturdy and can take the weight of these cheese blocks for a prolonged period. Also, these should be placed at a proper distance for comfortable handling and moving of your cheese blocks. It must also provide optimum air circulation and movement. To regulate moisture levels, use a humidifier.

All in all, take care of these considerations to make an effective cheese cave in your backyard.

Temperature control. A steady temperature between 45 and 55 F is ideal for your cheese cave. The temperature can be controlled using the direction of your cave (to prevent sunlight), frost line, direction, and types of openings, the heat generated and release from lights and cheese, and the level of insulation. You can also use artificial cooling and heating systems to regulate the internal temperature.

Humidity. We need relative humidity between 85 and 95 percent, which can be achieved by humidifiers, the presence of cold water within the cave, and the level of permeability of the internal walls.

Air circulation. As cheese ages, it emits unwanted products such as carbon dioxide, ammonia, and other gases. These should be released and removed from the cheese cave as they can ruin the taste and smell of the cheese. For this, adequate air circulation is necessary and can be achieved by opening and closing the cave door every once in a while or using a device to bring in filtered air. Consider the amount and velocity of airflow, the quantity of cheese present, and the intake and exhaust of air within the internal environment.

Sanitation and maintenance. Lastly, make sure that you are keeping the cellar and cave clean as it could also affect the cheese's quality. Check the quality and components of the cleaning agents you are using, the location of the drains and washing area, local regulations, and regularly cleaning the shelves and equipment used to make the cheese.

Some Basic Recipes

Some very interesting recipes can be made with cheese. Let's take a look at some of them.

1. Caprese salad – slices of fresh mozzarella, tomatoes, and basil leaves topped with salt, pepper, and olive oil

2. Mozzarella sticks – Mozzarella cheese cuboids coated in flour and breadcrumbs and deep-fried.

3. Buttermilk cheese scones – Baked scones dipped in buttermilk and filled with cheese.

4. New York cheesecake – Use your fresh homemade cream cheese to make a baked cheesecake with a berry or dark chocolate sauce.

5. Quattro Formaggi Pizza – Fresh pizza baked with the bread recipe mentioned above and topped with four types of homemade cheeses of your choice.

Additional tip: Make a working kitchen that is easily accessible from your cheese making shed or working area, this is particularly useful when you are also growing fresh veggies and fruits in your backyard homestead. You can experiment with new cheese recipes and store them in your pantry. If you plan on making your cheese in the kitchen, make sure that you have a wide and comfortable space to work in as it can be a hectic process at times.

Chapter 5

Make your Own Drinkable Water

A functional backyard homestead is also capable of making drinkable water. You no longer have to pay for bulky plastic water bottles. You can follow specific water purification methods to make rainwater collection drinkable; this is a sustainable approach that saves a lot of money and resources. In this chapter, we will take a look at some of these water purification methods that work and successfully convert unfit water into drinking water.

Before we take a look at the water purification techniques, let's learn how to collect rainwater to make it a thoroughly sustainable and inexpensive approach. However, to collect and use rainwater, your location should have sufficient annual rainfall. If not, you need to rely on other sources of water as you have been doing, such as tap water. You can also consider up-cycling gray water.

Here is one simple yet effective rainwater collecting method that you can prepare in your backyard homestead and for daily use:

- Fit a gutter on the edge of your roof. Connect the end of the gutter with a pipe that is connected to the wall below. The pipe will carry the rainwater from the gutter to the ground. If you already have gutters on your roof, make sure that they are clean and free of leaves and dust. The same implies to your pipes too.

- Buy a big barrel or use an existing container to collect the water from the end of the pipe. Make sure that the barrel or container is thoroughly cleaned from the inside.

- Cut the downspout and place it directly in the container for it to receive the rainwater.

- Before you insert the downspout in the container, cut a small hole at the bottom of the container, and fit a tap to make the access easier. With this, you can fetch rainwater for purifying with ease.

- Place the container under a shed to regulate the temperature of the collected water.

Once you collect your rainwater, it is now time to purify it. You can use one or more of the following techniques to purify rainwater and make it fit for consumption.

1. Boiling Your Water

Boiling water is one of the easiest and most convenient ways to purify water and make it fit for drinking. It's quite simple too. All you have to do is pour the impure water in a pot for boiling and heat it until it reaches a rolling boil. Let it further boil for at least 10 to 15 minutes. Boil the water for as long as you can, as it will make it purer and safer for consumption. However, there is a disadvantage to boiling water. Boiling it will remove all the oxygen present, which will ruin its taste. If you are okay with flat-tasting water, boiling can be your primary water purifying technique. If not, you will get used to it eventually. To fix the taste, you can also shake the purified water to retain some of its taste. Lastly, adding a pinch of salt per quart of water will give an extra touch of flavor.

2. Distillation

Distillation is a process of collecting vapor by heating water and turning it back into a liquid for consumption. Compared to the contaminants and microbes present in water, the latter has a lower boiling point, which makes distillation an effective purifying technique. Water is boiled and left at its boiling point for a prolonged period. The converted vapor is then collected in a condenser and cooled down. The contaminants and harmful substances present in the water are left in the boiling container. Even though it is a slow process, it will remove all harmful bacteria, germs, and heavy metals such as arsenic, mercury, and lead. It also requires a reliable heat source.

To make your slow sand filtration system at home:

- Make a double boiler by adding 8 cups of water in a large pot and placing a smaller pot inside such that it floats. Make sure that there is enough space around the edges of the smaller pot for adequate air circulation.

- Place the large pot on medium-high flame and let the water simmer (not boil) at a temperature of 180 to 200 F.

- Once the water starts simmering, place a lid over the top of the large pot in an upside-down position. This will allow the vapor to trickle down the center of the lid and collect it in the smaller pot.

- Meanwhile, place some ice cubes over the top of your lid, it will accelerate the condensation process.

- After around 30 to 45 minutes, you will notice some water collected in the smaller pot, which is now distilled.

Distillation is a long and energy-consuming process but gives promising results.

3. Chlorination

This process uses chlorine as the main ingredient to purify water and is typically used in many households. It is a strong chemical that kills harmful pathogens, germs, and other organisms present in water that make it unfit for drinking. You can use liquid chlorine or chlorine tablets to purify water. Pop them in heated water (follow the instructions for the quantity) and let it dissolve when the water reaches around 70 F or higher. Certain individuals with health problems such as thyroid or related issues should take advice from their medical practitioners before delving into this method.

4. Slow Sand Filtration

As the name suggests, this method uses sand to filter contaminated water and is usually used by commercial farmers, making it an economical, sustainable, and effective water purification technique. In this method, you use different layers of sand that produce a gel-like substance on the surface, which filters the water and removes impurities. It's a great DIY setup that does not accommodate much space and can be easily prepared and installed in your backyard homestead. Moreover, these do not require any chemical and mechanical parts or support for its operation.

To make your slow sand filtration system at home:

- Cut a sanitized plastic bottle in half and turn it over so that it stands on its cap. Remove the cap and wrap the mouth with a coffee filter and secure it tightly with a rubber band.

- Prepare a holder for the plastic as it will stay in this position for a prolonged period.

- Prepare the first layer with fine sand and even it out. Next, pour coarse sand on its top and even it out - this is the second layer.

- For the third layer, place small pebbles and spread them evenly.

- The last and final layer will have rocks or huge gravel. Make sure that you clean and sanitize these pebbles and rocks before using them.

- Place a collecting pot under the bottle and pour contaminated water over it through a strainer. Let it filter and drip in the collecting pot.

5. Solar Disinfection

This method is a sustainable way of disinfecting your water. This system uses solar energy to purify water instead of keeping it under extreme heat, like boiling water. For this, you need to fill in a few clean plastic bottles three-quarters full of the contaminated water. Shake it for around 20 to 30 seconds and fill the remaining part of the bottle with the contaminated water. Close the caps and place the bottles under sunlight for 6 to 8 hours.

If boiling, distillation, or any other method does not work for you, you can use solar disinfection as it is extremely inexpensive.

Storing Purified Water

The best way to preserve drinking water is inspired by a traditional way used in rural areas of Asia, which is earthen pots. Many people still use this method to date. Earthen pots are made of red clay that has cooling properties and provides a distinctive flavor. By storing water in earthen pots, you can get cool water during summer. If you preserve water in an earthen pot for a few days, you can achieve a distinctive taste that is favored by many. Moreover, you don't need a refrigerator specifically for cool water. You can buy earthen pots of various sizes online or at any store that sells crafty vessels.

Another way to preserve huge quantities of drinking water is a tank that collects purified water directly from the system. Make sure that you sanitize and clean your tank regularly to prevent diseases and growth of microbes that could cause potential health threats.

Chapter 6

Make your Own Tea

Next on our list of making your own produce includes tea. Imagine sipping on a cup of hot tea on a rainy day while sitting in your pretty backyard homestead. It seems surreal, doesn't it? With a backyard homestead, you can prepare your preferred tea. Whether it's green, black, white, or oolong tea that you devour, you can now grow and make your tea at home with this simple guide.

Before we delve into the process of growing your tea, let's understand what tea actually is. By knowing your plant, you can understand its characteristics and achieve better results. The tea plant is Camellia Sinensis, or just C. Sinensis, from which different varieties of tea can be extracted. If left unpruned, your tea plant will grow up to a height of 20 to 30 feet tall. However, you should prune it regularly. Two different types of tea are inherent to the C. Sinensis tea plant, one of which hails from southern China (grown in cooler climates), and the other one is grown in Assam, India (one with larger leaves). There is another kind called Camellia assamica, which is mostly found in Africa.

Basic Setup

You can develop garden beds and fill them with well-drained and sandy soil for the best results. If your location is too cold, you might have to invest in movable teapots that can hold tall plants. If it is too cold outside, you can simply take your teapots inside. Homesteaders living in Zone 8 (which is in the Midwest and the southern parts of the USA) can easily grow various tea plants in their backyards due to the perfect weather conditions.

Preparing the Soil

You need to pay attention while preparing the soil because it will eventually decide the health of your tea shrubs. The preferable types of soil include loose or slightly sandy soil that drains well. If your soil holds too much water it won't do your shrubs much good. If you are unable to find loose or sandy soil, you can also go for loam and porous soil that has a higher level of acidity in it. However, make sure that the soil you choose has a lesser calcium content. Three factors you need to take care of to provide optimum growing conditions for your tea shrub are –

- Porous, loose, or sandy soil

- Lower calcium level

- Soil with a higher acidity level

Once you have your hands on the soil mixture, it is time to transfer it to the garden pot, where you will be growing your tea shrubs. Add a small amount of sphagnum moss to your soil to increase its nutritional quality.

Choosing and Planting the Tea Shrub

When your soil and planter is ready, it is now time to plant your tea shrubs in it. You can buy an actual tea plant from your nearby nursery or go for a variety of teas based on its different parts, such as herbs, flowers, roots, or stems, etc.

You can grow one or all of these seeds and plants from which you can extract different varieties of tea –

1. Camellia Sinensis

As we know by now, this is the truest tea shrub that will produce the actual tea with its original essence. You can choose between large-leaved and small-leaved tea plants based on the availability and weather conditions in your area. It is easy to manage since it grows up to a height of just 3 to 6 feet when pruned regularly. These shrubs are aesthetic to look at and turn into ornamental plants in the fall season. You can buy seeds or seedlings of Camellia Sinensis at your nearest nursery. As they grow, they resemble a bush, but if not managed and pruned properly, they might start looking like a tree. You need a lot of patience when growing tea, as this plant can take up to 3 years to fully mature. As soon as these grow, use a mortar and pestle to crush the young buds or press them between your hands to make them drinkable. With this plant, you will get white, black, green, and oolong tea.

2. Rosehip

Rose hips are extracted and grown from the seed pods of wild roses and are deemed as berries. The taste range of rose hips varies depending on the type of seed pod it has grown from. You can expect rosehip tea leaves that taste too bland or super sweet. Additionally, the tea that grows from this plant is highly nutritious and contains essential nutrients such as Vitamin A, E, and C.

Some people use tea leaves from a garden rose plant leaves. However, these are usually sprayed with chemicals and shouldn't be used. Rosehip plants are ideally grown in sunny locations, so if your location is hot almost all round the year, this plant should be your primary choice. Well-drained soil is the best choice for this plant. Even if you don't have the apt soil or location, you can grow the plant in a pot and optimize its surroundings for the best results. Pluck dried or fresh rose hips to make your tea. However, be careful while harvesting tea from this plant as it is usually full of thorns. Since rose hips contain zero caffeine, the tea from this plant is not bitter. You can still strengthen its taste by increasing the number of rose hips.

Follow these steps to use rose hips for making tea –

- Select the youngest leaves and buds and crush them between your palms.

- Pour some water in a pot and heat it until it reaches a temperature of 190 F.

- Add the crushed rosehip leaves in this heated water and let them cook for a while.

- Strain it in a cup and add honey or lemon to serve.

Apart from being tasty and refreshing, rosehip tea also improves your health. It contains 20 times Vitamin C content as in oranges, which is excellent for your immune health.

3. Dandelion Root

This type of plant is usually considered a waste in one's lawn. However, the roots of this plant can make for a delicious and healthy cup of tea. It contains a lot of antioxidants that remove toxins from one's body and cleanses it from the inside. It further helps in strengthening the immune system and reducing the chances of forming tumors and is also known to promote a healthy liver. Moreover, this tea is great for people suffering from urinary tract infection. Dandelion roots can be grown in spring and fall. However, there are multiple arguments claiming the ideal season for growing dandelion roots. It is found that summer isn't the best time as its growth focuses more on growing and nurturing the flowers and leaves.

Be extra careful while harvesting dandelion roots for making tea. Use your hands to pull out the leaves and use a fork to dig out the roots. Take the crown of the plant as a reference and dig out three inches of space around it. Once you are able to extract it, remove the top of the plant and detach it from its roots by cutting it. Instead of throwing away the top part of the plant, you can use it in tinctures and salads.

To use the roots for tea:

- Wash it properly to get rid of the soil. Scrub the roots using a vegetable brush to clean it thoroughly and remove buried soil.

- Once these are clean, cut them up in equal slices (almost ¼ inch long).

- Use a dehydrator or pop these slices in the oven to dry them.

- Remove from the oven and let them cool at room temperature for a few hours.

- These are ready to be used. Transfer in an airtight container and place it in a dry and cool spot. You can use these for up to 1 year.

To prepare tea for drinking:

- Pour 2 cups of water in a pot and heat until it reaches a temperature of 190 F.

- Add 2 tsp of dried dandelion root tea slices in this heated water and let them cook for about 15 minutes.

- Strain it in a cup and add honey or lemon to serve.

4. Coriander Seeds

Coriander is often confused with cilantro. Even though these two come from the same plant, they have different properties and taste. The entire plant is known as coriander, and the leaves are cilantro. You can grow coriander from seeds or through starter plants. Coriander grows better in milder climates, such as spring and fall. It shouldn't be kept under direct sunlight. This plant is easy to grow and manage if tended to every day.

You need coriander seeds to be used in a tea. To harvest these, cut the seed heads that are still green and place them in a paper bag to dry. You can also harvest dry seeds directly from the plant. You can either keep them in an airtight container for frequent use or freeze them for prolonged use.

To make drinkable tea from coriander seeds:

- Pour 2 cups of water in a pot and heat until it reaches a temperature of 190 F.

- Add a few coriander seeds in this heated water and let it cook for about 15 minutes. You can also pour this heated water directly over some coriander seeds. Let it steep for around 5 to 7 minutes.

- Strain it in a cup and add honey or lemon to serve.

Just like other teas, the tea made from coriander seeds is known to prevent digestive issues and cure other stomach related ailments such as food poisoning.

5. *Lavender*

This is one of the most aromatic and tastiest tea types among all. Because of its pleasant aroma, it is used to calm nerves and is one of the main treatment routes in aromatherapy. It offers a calming effect and helps reduce anxiety and increase concentration. Also, lavender plants are packed with nutrients such as Vitamin A, Vitamin C, calcium, and essential amino acids. It is excellent for boosting your immune system, too. The taste and distinctive smell of this plant are due to the presence of camphor in its flowers, and it enhances the sweetness of the tea.

Lavender is another easy and carefree plant that can be grown in the comfort of your backyard. You can grow this in a pot or directly plant it in your backyard garden. Make sure that the soil is well-draining, and the pots are placed under direct sunlight. Instead of growing it from scratch, it is wiser for novices to grow it from lavender starter plants. Keep an eye on your lavender plant and prune it every once in a while for healthy growth. During summertime, multiple lavender buds will bloom, which can be plucked and used for making tea. To harvest the buds for tea, just pluck them, wash them, and store them in an airtight container.

To make drinkable tea from lavender buds:

- Pour 2 cups of water in a pot and heat until it reaches a temperature of 190 F.

- Add 2 tsp of fresh lavender buds in this heated water and let it cook for about 15 minutes. You can also pour this heated water directly over the lavender buds. Let it steep for around 5 to 7 minutes.

- Strain it in a cup and add honey or lemon to serve. You can also add other refreshing ingredients such as mint, chamomile, or berries to enhance the taste.

If you are constantly facing health issues, such as coughs and colds or an upset stomach, you should drink a cup of lavender tea every day.

Process

For the sake of better understanding, we will divide the process into two parts – growing tea and harvesting and processing it.

Growing Tea

As mentioned, you can prepare your soil and check your local guidelines for apt weather conditions. Decide the type of tea you want to grow and prepare the soil accordingly. By now, you have a fair idea of how the ideal soil should be. Prepare the pots, plant the seeds or plant, and water it frequently; this step needs some patience and constant observation. Depending on the type of plant you have growing, it will take a period between 6 months to 3 years for the tea plant or parts to grow.

Harvesting and Processing

When your tea plant has fully grown, it is time to harvest and process it. Follow this step by step guide to harvest and process five types of teas, which are green, black, white, oolong, and rooibos.

1. Green Tea

- Select the youngest leaves and buds and place them on a flat surface with blotting paper.

- Dry them thoroughly with the blotting paper and leave them in the shade to air dry.

- Place a pot on the stove and pour in water to heat. Place your leaves on a steamer and steam them for around 60 to 80 seconds. If you want a nice, toasted flavor, skip the steaming and roast them in a skillet instead for the same duration.

- Meanwhile, prepare your baking sheet. Once the tea leaves are roasted or steamed, place them on the baking sheet and place them in the oven at a temperature of 250 F. Let them dry bake for around 20 minutes.

- Once done, remove from the oven and let them cool. Transfer into an airtight container and place it in a dry and cool spot.

2. Black Tea

- Select the youngest leaves and buds and crush them between your palms until they form a black color. As the leaves are rolled and crushed, you will notice that the edges will turn red.

- Prepare a tray and spread the crushed leaves. Leave them to dry in the shade for at least 2 to 3 days.

- Meanwhile, prepare your baking sheet. Once the tea leaves are dry, place them on the baking sheet and place them in the oven at a temperature of 250 F. Let them dry bake for around 20 minutes.

- Once done, remove from the oven and let them cool. Transfer into an airtight container and place it in a dry and cool spot.

3. Oolong Tea

- Select the youngest leaves and buds and place them on a flat surface, on a towel in the sun for around 45 minutes. You will notice that the leaves have wilted after 45 minutes.

- Place them indoors and let them cool down for a while. Stir the leaves every once in a while for even cooling.

- As the leaves will dry and cool, you will notice that the edges will turn red.

- Meanwhile, prepare your baking sheet. Once the tea leaves are dry, place them on the baking sheet and place them in the oven at a temperature of 250 F. Let them dry bake for around 20 minutes.

- Once done, remove from the oven and let them cool. Transfer into an airtight container and place it in a dry and cool spot.

4. White Tea

- Select the youngest leaves and buds and place them on a flat surface with a towel in a shaded area for around 72 hours. You will notice that the leaves have wilted after this process.

- Place them indoors and let them cool down for a while. Stir the leaves every once in a while for even cooling. For better results, dry these leaves using hot air blasts or through steaming. The key is to process it for a lower period compared to black tea as it is milder in flavor.

- Transfer into an airtight container and place it in a dry and cool spot.

5. Rooibos

- Select the youngest leaves and buds and allow them to ferment for a while until the edges turn red.

- Pile the stems, leaves, and buds to allow faster fermenting. Upon fermenting, you will notice that the entire pile has turned red.

- Place the fermented tea leaves on a tray and spread them evenly. Let them dry for a while.

- Transfer into an airtight container and place it in a dry and cool spot.

Basic Questions

1. How Much Tea Can I Expect to Harvest in a Season?

You need to take good care of your tea plants at the beginning with adequate pruning and shaping. By doing this, you can expect a light harvest in the second season. If you continue to take care of your plants, you can expect a heavy harvest in the fourth or fifth season. It will also depend on the type of soil and climate in your location. If the plants are mature, you can expect a harvest of approximately one-half of a pound of leaf per plant. If you are a beginner, you can expect the harvest to be around one-quarter of a pound of leaf per plant. However, it is still enough to sustain your daily needs.

2. How Much Tea Can I Process After Harvesting It?

With proper practice and processing equipment, you can expect every pound of raw leaf to be processed into one-fifth of a pound of finished tea. In simpler words, every five pounds of raw tea leaves can be converted into one pound of processed tea. The loss is due to the evaporation of water that takes place during the processing stages.

3. What Kind of Weather and Climate is Apt for Growing Tea?

Since tea needs strict environmental conditions and climate to grow, you should consider the climate of your location before you decide to grow your tea. Tea plants cannot survive extremely cold weather, which is below 0 F and up to 15 F. If you live in a cold location, one harsh winter will ruin all your year-round efforts of growing and tending to your tea plants. To confirm the climatic conditions and the plant behavior in your area, you can check the USDA plant guidelines to determine your tea shrub's behavior. If you don't live in any of these suggested zones, you can still grow your tea shrub in a greenhouse.

The best soil for growing tea shrubs is slightly higher on the acidic level and very sandy. Your soil must also be well-drained.

Recipes

As we know, the best way to drink tea is to add a dash of lemon and ginger. It is a simple and healthy recipe that refreshes your palate and keeps away cravings. You can buy tea bags and fill them up with dried tea or simply use a tea infuser to make tea in hot water. Add other interesting ingredients such as hibiscus flowers or jasmine in summers to provide a refreshing taste.

However, there are many more recipes that you can make using tea, some of which include the produce from your backyard homestead.

Simple Iced Tea

Iced tea is one of the most savored drinks during summer days, and with your home-produced tea, it will taste even more delicious. It's a tasty and simple recipe that will barely take a few minutes to prepare. Making iced tea is also a simple and encouraging way to begin. It's also a great and healthier replacement of the bottled iced tea that contains a lot of sugar. It is a cheaper, healthier, and self-sufficient alternative that can be made at any hour of the day.

Equipment: A pitcher, a heatproof canning jar or boiling pot, and serving glasses.

Ingredients:

To make ½ gallon of iced tea - 6 tea bags or 6 tbsps of loose leaf tea per 8 cups of water

To make 1 gallon of iced tea - 12 tea bags or 12 tbsps of loose leaf tea per 8 cups of water

To change the measurements as per requirement, use 2 tbsps loose leaf tea per 2 cups of water.

- ½ lemon

- Ice cubes

- Sugar (optional)

- Mint leaves (for garnish, optional)

Directions:

1. Pour the water in the heatproof canning jar or boiling pot and bring it to boil. Change the measurements of the water to be boiled according to your requirement.

2. Add the proportionate amount of tea leaves or bags (as explained above) to the boiled water. Let it steep for around 15 to 20 minutes.

3. Add sugar if you want it to be sweet. You can also replace it with healthier options such as honey, maple syrup, or brown sugar.

4. Strain the mixture. Let it cool until it reaches room temperature.

5. Meanwhile, prepare your pitcher by adding ice cubes. Once the mixture has cooled, squeeze in some lemon juice.

6. Pour the mixture in the pitcher, and it is ready to be served. Use mint leaves to garnish.

You can prepare large batches of iced tea and preserve it in your refrigerator with a lid on your pitchers. However, we'd recommend you prepare fresh iced tea whenever you crave some as it tastes better, is easy to make, and hardly takes a few minutes.

Conclusion

By following these steps, you are now ready to build your backyard homestead and make your produce. It's a great way to step out of debt and build a self-sufficient lifestyle. It might sound overwhelming in the beginning, but once you get the hang of it, there's no turning back. Don't worry if you don't have a green thumb; just keep on practicing and follow the procedure with precision. The rest will fall into place. It's a craft that needs consistency, care, patience, and practice to master. Just start, learn on your way, and keep doing it without a second thought. Even if things don't work out in the beginning, do not stop. Backyard homesteading isn't child's play; one is bound to fail without any prior experience. However, do not stop merely because of that one tiny failure. Most successful backyard homestead owners have failed in the beginning, but they did not stop. You can do it too. Imagine the results in the end and stay motivated. Talk to fellow backyard homestead owners to gain inspiration.

As we have been emphasizing the importance of starting small, you should pay extra attention to it. Don't jump on projects that seem unattainable. It is wiser to finish a small project with successful results than starting a big project and leaving it incomplete. Take one step at a time and BE PATIENT. Learn from your mistakes and keep going.

Make proper use of that backyard of yours. Study and prepare your backyard today to build your homestead. You are just one step away. Good luck and happy homesteading!

References

https://thehouseandhomestead.com/easy-no-knead-homemade-bread/

http://homesteadinghomemaker.blogspot.com/2020/01/bread-making-101-family-homestead.html

https://www.bbcgoodfood.com/recipes/focaccia

https://tastykitchen.com/recipes/breads/honey-herb-bread/

https://butterwithasideofbread.com/peaches-and-cream-bread/#wprm-recipe-container-34580

https://www.thekitchn.com/whats-the-difference-between-active-dry-yeast-and-instant-yeast-54252

https://www.thekitchn.com/bread-baking-clinic-under-knea-157484

https://www.thekitchn.com/longer-lasting-yeast-store-dry-yeast-in-the-freezer-179315

https://www.thekitchn.com/active-dry-instant-yeast-best-tips-for-working-with-yeast-180312

https://www.thekitchn.com/fresh-baked-how-to-tell-when-b-106715

https://www.finecooking.com/article/the-science-of-baking-with-yeast-2#:~:text=As%20bread%20dough%20is%20mixed,causing%20the%20bread%20to%20rise.

https://cheesemaking.com/blogs/learn/how-to-select-cheese-wrap

https://cheesemaking.com/blogs/learn/finding-good-milk-for-cheese-making

https://cheesemaking.com/blogs/learn/how-to-make-a-cheese-cave

https://cheesemaking.com/blogs/learn/how-to-wax-cheese

https://cheesemaking.com/blogs/learn/equipment-for-cheese-making

https://cheesemaking.com/blogs/learn/ingredients-for-cheese-making

https://www.thespruceeats.com/30-minute-fresh-mozzarella-cheese-recipe-1806489

https://www.thespruceeats.com/homemade-mascarpone-cheese-recipe-1806499

https://cheesemaking.com/blogs/learn/the-cheese-making-process

https://www.thespruceeats.com/homemade-ricotta-cheese-recipe-591554

https://www.culturesforhealth.com/learn/recipe/cheese-recipes/cheddar-cheese-recipe/

https://www.thespruceeats.com/growing-tea-at-home-766090#:~:text=For%20planting%2C%20Camellia%20sinensis%20likes,before%20you%20start%20harvesting%20leaves.

https://www.johnquinnrealestate.com/growing-tea-at-home/

https://www.cnet.com/how-to/how-to-make-distilled-water-at-home-for-free/

https://www.schultzsoftwater.com/blog/4-methods-to-purify-your-water

https://mikesbackyardnursery.com/2013/07/8-ways-to-purify-water-at-home/

https://homesteading.com/best-homesteading-tools/

https://purelivingforlife.com/homestead-tools/j

BACKYARD HOMESTEADING

Growing Vegetables, Fruits, and Raising Livestock in an Urban House

MONA GREENY

Introduction

B ackyard homesteading is quickly becoming popular among people in urban areas. Still, the concept of it has been around for quite a while. Although it is widely known as backyard homesteading, you may also refer to it as backyard farming, urban homesteading, or suburban homesteading. But no matter what you call it, the basics of backyard homesteading is quite straightforward – take the space in your home and start growing food and raising livestock to live a sufficient life.

One thing I find interesting about backyard homesteading is that it became something like a trend around 2014. Many people consider it the next trendy thing to jump on as if it were a revolutionary discovery. However, the fact is that backyard homesteading is a concept that has been an integral part of the human experience in another way. Look back to the great depression when people had to take rabbits for their meat, and you will realize just how much backyard homesteading has been a part of the tried and authentic life.

The idea of backyard or urban homesteading is to help you start living a self-sufficient life – one where you can provide your family with fresh food directly from the source. So, if you are ready to take the plunge but you have no idea where to start, this book was written for you.

Backyard Homesteading contains all the information, guidelines, and tips you need to start your own backyard farm and start getting fresh food directly from the source. Necessarily, there are five steps that you have to take to get started with backyard homesteading. The first step is to start a garden; the second is to start composting; the third is to raise chickens, specifically for their eggs; the fourth is to raise other animals to serve as your meat source; and finally, the fifth step is to bring the mix from your backyard into your home. I know you are wondering what the last part means, but all will be revealed in the book.

It may be tempting to consider backyard homesteading a new trend to hop on, but you shouldn't. If you go into the practice with the mindset that this is just another internet thing you need to participate in, to feel like a part of the online community, you are more than likely to fail at backyard homesteading. Backyard homesteading is meant to be integrated into your life so that it becomes a part of you and your family – you need to take it seriously. Therefore, if you are someone who is looking for another trend to jump on, I suggest you move on to another book – this is a serious business. However, if you are looking for a book that will teach you to live a genuinely self-sufficient life, then this is the perfect book for you.

This book covers everything you need to know about backyard homesteading, from growing fruits and leafy greens to raising livestock for meat and eggs. In the first part, we look at the several benefits of backyard homesteading to help you fully assimilate the level of disservice you would be doing yourself and your family if you see backyard homesteading as just another internet trend. The first part also concentrates on how you can plan and start your own home garden for fruit and vegetable planting. Other parts of the book talk about tips, techniques, and methods of farming, raising livestock, as well as harvest and preservation of the foods you grow. Overall, this book provides detailed guidelines on every step you have to go through to jumpstart your backyard farm in your urban home. It promises to be an informative and educative read for people who want a comprehensive guide on how they can live a self-sufficient life.

So, if you are ready to start producing all the food you need in that little space at the back of your apartment, take a seat in your favorite chair to get started!

PART I

Backyard Homesteading

In the most straightforward language, homesteading refers to a lifestyle of self-reliance and self-sufficiency. Although this book will be focusing on the aspect of homesteading that has to do with subsistence agriculture, it actually involves everything from subsistence farming to food preservation and household craft-making. According to Wikipedia, homesteading is, "a lifestyle of self-sufficiency, which is characterized by subsistence agriculture, home preservation of foodstuffs, and may or may not involve the small-scale production of textiles, clothing, and craftwork for household use."

Homesteading has become more popular among Americans as a result of the issues affecting food security and health and wellness. Many people are finding ways to become more self-sufficient due to these issues, and homesteading just happens to offer the perfect solution. Depending on who you pose the question to, you might get different definitions of what homesteading means. Homesteading may mean something to you, and something entirely different to another person. It all depends on what you are willing to integrate into your lifestyle. But really, homesteading is basically about self-sufficiency, and you can practice it no matter where you are.

In the mainstream media today, though, homesteading is regarded as an act of keeping an extensive fruit and vegetable garden while also raising livestock to complement and ultimately provide enough food for family consumption. Some people often decide to sell part of their produce gotten directly from their backyard garden. Basically, people who have made backyard homesteading a lifestyle produce everything they need to survive food-wise themselves. As homesteading becomes more and more popular among people, many families in the suburbs are learning to start their own little farms and even dabble in animal husbandry. This is primarily why we say the concept of backyard homesteading teaches people to live a life of self-sufficiency even in their urban homes.

Although you may have had some people try to convince you that homesteading is really easy, you should know that it is demanding if you want to take it on full-scale. If you plant large-scale without ensuring that you can handle the demands of being an urban homesteader, it is doubtful that you will succeed at your homesteading venture. This is why you mustn't see backyard homesteading as just another trend to be a participant in; you must consider it a serious business – something to ensure your survival and improve your wellbeing. The best thing is to start slowly to avoid becoming overwhelmed by the activities and tasks that come with being a homesteader. Even if you reside in an apartment, there are tips you can use to get started with easy, hands-on activities before you move on to more challenging aspects of homesteading when you move to a new home.

The most comfortable and most stress-free way to start backyard homesteading is to begin producing some of the food you eat. You can start by planting an inexpensive vegetable garden, and ensure that your family members take part when it is time to harvest the produce. If you live in an apartment, you can quickly hone your homesteading skills. This will help you understand what it is like before you get your own property and become a full-blown homesteader.

Now, when it comes to becoming a micro homesteader in an urban home, people always have different excuses as to why they just can't start homesteading. Personally, it took me a couple of years before I had the balls to start the journey. There are many reasons why people opt to pass up on homesteading opportunities in their homes. Some people are understandably afraid of making mistakes. In contrast, others are just scared of being the "weird" one in the neighborhood. For many other people, the mere thought of becoming a micro homesteader is just overwhelming. This usually makes them procrastinate and then give up altogether. However, apart from these reasons, there are other reasons why many people still think that homesteading isn't such an ideal thing for urban residents.

The primary reason is due to the many popular misconceptions about homesteading. In most cases, some of these misconceptions are outrightly false, while others are misunderstood. You have likely heard at least one of these misconceptions and probably even believed some of them. It is essential to correct misconceptions and myths as they are the reason why many people are still deterred from the idea of starting a micro homestead garden and livestock farm.

One of the most widespread misconceptions about homesteading is that you need tons of land to do it. As a result, many people don't believe that you can start a farm or raise animals in an urban home. Some people even think that you need to live in a rural area if you want to practice homesteading. In the past, Americans equated homesteading to life in the country. They believe that you can't own a garden unless you live in the countryside. However, the truth is that you don't need tons of land spaces to be self-sufficient – you only need to be creative with how you use the little space you have. Whether you live in a rural or suburban area, you can start backyard homesteading wherever you are.

According to a study published by some researchers at the University of London, England, an acre of a suburban garden can actually provide triple the size of produce you would get from an acre of farmland. This is because a family can quickly become intimate with the one-acre they have and devise different ways to navigate around it. When you have just one acre of land to grow your garden on, you are sure to quickly know the needs of that garden and the best ways you can manage it to get your desired results. Some people start their homesteading lifestyle with chicken-rearing; others start a vegetable garden first. It does not matter what you start with – the point is to begin providing food for yourself. As long as you are making your food from scratch and raising and growing food on your own mini urban farm, you are self-sufficient.

Another misconception people have about urban homesteading is that it ruins the look of their house. As a result, many people think you should only grow food in your background. Well, growing food in your home can be much more creative than merely planting a plain-looking square garden in rows. And, you don't even need to confine your garden to the backyard. Fruit gardens can be grown out in the front yard, which will add to the beauty of your home. With the rows of beautiful fruits, flowers, and leaves, nobody can say that homesteading ruins the look of a house. As a matter of fact, the edible landscape adds to the beauty of the home. Undoubtedly, there are always two ways of doing anything – the high-maintenance or low-maintenance way. The approach you take depends on your budget and the amount you have to invest in your homesteading project.

If you live in the suburbs, my bet is you didn't think of farming when you were choosing your property. How many suburbanites select a property for its farming advantage anyway? As a result, most suburbanites have properties that pose specific challenges to the farming lifestyle. For instance, some properties have sloping land. At the same time, there are too many shady areas in some, making it impossible for the sun to really shine through. While most people think that sloping land is a disadvantage that makes it impossible to grow crops, the truth is that sloping or hilly lands present you an opportunity to grow a variety of plants. Also, a yard with shade is excellent for producing a diversity of vegetables, including leafy greens and root vegetables. So, the fact that you live in the suburbs isn't really the disadvantage that many people think it is. Instead, it is an opportunity you can take advantage of to produce an abundance of produce every year.

One very interesting misconception about micro or backyard homesteading is that you can't grow beyond vegetables and fruits if you want to be a backyard homesteader. Of course, there is no lie in saying that the more time you invest in making something work, the more returns you get out of it. This also applies to backyard gardening and homesteading. However, it does not mean that you can only grow vegetables and fruits. Space is the only thing that can hold you back from expanding your backyard garden to include things other than fruits and vegetables. You must be realistic about the space you have for your homesteading journey, but you can surely start with any space, no matter how small. To get the most out of your garden, you have to make gardening an integral part of your daily routine. Even if it is just 15 minutes, dedicate it to your garden each day. This makes it easy for you to stick to the method, also when you feel like not doing anything on a particular day. If you don't start by investing a few minutes and you keep waiting for the day, you will have a row of uninterrupted hours that that day may never come. The little steps you take are what will eventually take you to your goals.

For whatever reason, a lot of people believe that backyard homesteading isn't homesteading until you add some farm animals to your micro-farm. This is false. If you want to be a traditional farmer to the core, feel free. However, don't feel compelled to include farm animals in your homesteading plan unless you really want to. The point of backyard homesteading is so that you can grow the crops you wish to and produce the food you eat yourself. If you don't want to raise livestock, you can just stick to fruit and vegetable gardening. But if you would like to raise livestock as egg and meat sources, know that you can also do that. It is all about wants, preferences, and choices.

There are many approaches you can take to backyard homesteading. But no matter the method you want to use, understand that you can start wherever you are, right now. As long as you know how to go about backyard homesteading, you don't have to procrastinate or shy away. Also, it helps to transition slowly when you start. You don't have to start cold turkey or go into it entirely at once. If you don't want to commit to that lifestyle fully just yet, you don't have to do it – just take everything one step at a time. If you can't handle tending to many different fruits and vegetables every day, you can start by growing just one or two simple vegetables or fruits and gradually build up.

The good thing about starting backyard homesteading is that you can get your inspiration from those who have made it a lifestyle. There are hundreds of thousands of people who have successfully started and incorporated homesteading into their lives. Also, you can quickly begin to wherever you are and whenever you choose. You do not have to wait to hone any specific skills; just acquire the basics, and you will be good to go. Even if the space you have is only on your balcony, you can start gardening there with the right knowledge and tips.

The next chapter elaborates on the benefits of backyard homesteading in an attempt to help you understand just why homesteading is right for you.

Chapter One

The Benefits of Backyard Homesteading

If you ask any homesteader how homesteading makes them feel, they will likely tell you that they feel blessed. And the truth is you won't know just how fortunate they feel until you become a homesteader yourself. You start reaping the benefits of getting your food directly from your backyard farm. The many benefits of backyard homesteading are enough reasons for you to start backyard homesteading if you are still on the fence. Undoubtedly, making a transition from getting your food from the grocery store to getting it from your backyard farm isn't always easy for homesteading beginners. However, it does get better with time.

And then, there is the whole thing with convincing your family about the new vision and goals you have. Most people find it challenging to convince their families of how beneficial their new lifestyle would be once they get started. In this age, it is effortless to come up with different excuses as to why homesteading may not be the right lifestyle for you. For instance, some people don't want to be homesteaders because they think it'll make people see them as hippies. Others simply don't want the inconvenience that comes with being a backyard homesteader, even though the lifestyle is far from being inconvenient.

"Why start a micro-urban farm when you can just get your food from the grocery store?" Well, I can assure you that sourcing the food you consume by yourself, on your own farm, is much more satisfying and helpful than buying stale food from your local grocery store. The best time to begin working on your backyard homesteading plan is NOW, and you shouldn't let anything deter you. Even if you have to start by taking baby steps. And if you experience setbacks, if people will think you're strange because you have decided to be an urban farmer, know that it always ends up being worth it. So, if you need some extra push to start working towards your goals, here are some benefits of backyard homesteading that should convince your family and even you.

Food Awareness

Unsurprisingly, many people in society are unaware of where food comes from and how it arrives at the dinner table. Children, in particular, don't have the slightest hint or clue of where their favorite meals come from. Homesteading is the key to educating your children and your family about food and where it comes from. A micro-urban farm encourages you to develop an intimate connection with the cycle of food production. This knowledge is something every human, no matter how young or old, should have. It helps you understand and appreciate the seemingly trivial achievement of being able to put food on your table. Something is satisfying about knowing where your food comes from. It helps you understand nature more.

In a way, the homesteading experience is also humbling. Homesteaders are usually quick to understand just how finite life is. As a beginner to urban homesteading, you will make a lot of mistakes. These mistakes can be overwhelming. Your livestock will die. Seemingly healthy crops will also die. Structures may collapse, and your plan may fail more than once. The experience is humbling for many people. If you keep chickens, they will likely be attacked and eaten by predators. But, regardless of the failures and lessons, you are going to continue homesteading if you are really keen on it. If anybody tells you there is a way to practice backyard homesteading without making mistakes at all, you should know that person is playing on your naivety. So, as homesteading is a quite humbling journey for people, it helps build perseverance. Urban homesteading is character building.

With backyard homesteading comes a level of freedom that most people haven't achieved in their lifetime. Due to the self-sufficient lifestyle, many homesteaders tend to become relatively independent, usually more than they have ever been. Becoming an urban homesteader frees you from the centralized food supply. Most homesteaders don't worry about people complaining about the inflation of dairy products in the market. If you have a cow, why do you have to worry about the rising price of milk? Even if beef becomes more expensive, homesteaders don't worry because they know they have their livestock. The increased level of freedom from price-hike at the market makes your heart giddy and happy. It is enough reason to become an urban homesteader today.

Security

To some extent, homesteading offers security during extreme times. It does not matter whether your issue is a small or significant concern; you can always count on homesteading to provide a level of security in terms of foods and skills. If you know any homesteader, then you probably know that they still have a supply of food on hand because: 1) when you grow your food, there is always extra to preserve and store away. 2) Many homesteaders cannot help keeping mason jars and canning supplies. Although your personal food preservations techniques may need a little polishing when you become a homesteader, the fact is that you will always have enough food supply to last you for months, in the pantry, cupboards, basement, and freezer. Plus, some of the skills you hone from becoming a homesteader can be really helpful in extreme survival scenarios.

Work Ethic

Unsurprisingly, homesteading also helps to sharpen your work ethic. You'll agree that one thing that is currently missing in world culture today is a strong work ethic. Go back to the times of our ancestors, and you will find that children started learning all about providing their own food, milking cows and growing crops from the ages of six and seven. In those times, children already knew how to feed animals and train oxen. Nowadays, though, the environment is entirely different. The truth is that children become capable, and they thrive in any situation that encourages them to partake in worthwhile activities. While you should be grateful that your children don't have to go through such intense labor in this age (thanks to technology and advancement), you can't deny that homesteading can help build up your kids with strong work ethics. It doesn't matter how little they do. If there is anything that can help children build themselves up to develop a strong work ethic, it is the responsibility that accompanies growing food. It teaches them and you so much and it comes with the responsibility of ensuring that everyone can eat and survive; and the livestock you raise depends on you for their own survival, day after day.

Healthy, Tasty Food

Food that comes directly from your own farm tastes better than food from the grocery store. Eggs that come from your own poultry looks a lot healthier and taste better than store-bought eggs, for good reasons. If you know anyone who keeps chickens, then you should know how bright and beautiful the yolk from healthy backyard hens are. The taste is always incomparable. The depth of the difference between homegrown food and conventional food is simply incredible. Homesteading food tastes good. Homesteading also teaches you to be appreciative of what you have. When you understand the level of work that goes into growing your own food – from planting the seeds to caring for the seedlings and nurturing them until they mature – it is hard not to show appreciation and be grateful. Nothing nurtures gratitude and satisfaction more than knowing the amount of hard work that goes into providing your family and loved ones with food to eat. Inadvertently, this also teaches you to treat your crops better and appreciate their values.

There are many more benefits of backyard homesteading, but this should be about enough to convince you and anyone else you would like to convince on the reasons why backyard homesteading is ideal for your family. In the next chapter, we discuss how you can get started with backyard homesteading. What is the first thing you do to start homesteading in your urban home? Let's find out in the subsequent chapter.

Chapter Two

Planning and Getting Started

Now that you understand why you should become an urban homesteader; you may be wondering where exactly to start from. Figuring out where to begin homesteading can be pretty challenging, especially when you have zero background knowledge about farming or owning land. The point of this chapter is to help you demystify the whole process and give insight into the concrete steps you should take to begin your homesteading journey.

Before you get started with homesteading, though, you should be convinced that it is a lifestyle that you really want. Many people have this idealized perception of what homesteading should be like in their head. Still, it often turns out, unlike what they imagined in their head. Seriously, you should take some time to reflect on the daily activities, tasks, and chores that you will have to complete as a homesteader. Caring for crops and livestock is much harder than most people think. The activities are time-consuming and physically demanding, and not everyone has the stomach for that. This is why it is essential to ensure that your partner is also on board before you start backyard homesteading. Make sure it is the kind of lifestyle they want. Sit down and have an open discussion about your homesteading goals. If your partner doesn't share the same sentiments as you, then you will find it challenging to live a homesteading lifestyle.

The first place to start is to pick the aspects of homesteading that you want to focus on. This means that you have to define your priorities. Pick one or two homesteading projects that you are confident that you can start and commit to in the following months. If you are a suburban resident, you may want to consider starting with chicken-keeping and fruit or vegetable garden. For instance, you can get 3 to 4 chickens to keep for eggs. There are steps you will take if you want to focus on keeping chickens, and you will find these out as you read further. This book focuses on how you can start your own fruit garden, vegetable garden, and also start raising livestock. However, you don't have to do all these things at once. You can begin by growing vegetables if that is what you can fully commit to at the moment. Then, as you get used to the homesteading life, you can expand to growing fruits and raising livestock.

Planning and setting goals are essential to the homesteading journey. Without a formulated plan, you are unlikely to succeed at your homesteading venture. Making a plan for each step of your backyard homesteading adventure is vital to achieving the specific goals you have set for yourself. One thing about backyard homesteading is that you don't have to move to the country or have your own property. Backyard homesteading encourages you to live a sustainable life by growing a garden, raising some livestock, and preserving the food you make. So, the goals you make have to align with your setting. Your plan must include the homestead buildings, crops, and livestock, as well as how to make income if you want to. You can create your project on your phone or get a journal.

Start by listing the things you want to accomplish on your homesteading journey. What kind of homesteader do you want to be? Do you want to be off-grid ultimately? Do you want to produce all of the foods you consume each year without buying anything from the grocery store? Do you want to make income from your backyard farm? These are all questions that you must answer in your plan. The list you make will be crucial to your research. Research is a vital part of backyard homesteading. You have to research the best practices for the aspects of homesteading you want to focus on. This will help you make the best decisions when planning the phases of your homestead. So, look at the list you create and use it to plot out achievable goals. Once you set goals, it will be easier to take the next steps.

The goal you set will determine the size of the land you need. So, ensure that you keep this in mind as you make your plan and set achievable goals. If you want to keep your full-time or part-time job and just practice homesteading as a hobby, it is much easier to get by in an urban setting. But if you want to make homesteading your lifestyle, you will need enough space to grow all the fruits and vegetables that you need, plus space for the livestock you want to keep. Additionally, you will also need to establish parameters on the general area you want to live in. As a homesteader, do you want to live in a remote part of town? If you are planning to buy property to start your homesteading lifestyle, ensure the land you get will work for the homesteading lifestyle you want to live.

The next step is to conduct research. Before you start raising livestock in your back garden, you have to ensure that it is acceptable in the neighborhood you live in. For instance, keeping chickens requires you to check your local ordinances and ensure that you are allowed to keep chickens. You also have to check the maximum number of chickens you are allowed to keep to ensure you don't waste your money. The importance of research in backyard homesteading planning cannot be overstated. With proper research, you are less likely to face adversities. When you conduct in-depth and thorough research into something before you get started, you are less likely to be overwhelmed by the challenges you encounter along the way.

Budget is a critical part of planning. A significant portion of your plan will focus on setting a realistic and cost-effective budget that won't take too much out of your pocket. Creating a thoroughly researched budget is very important, especially if you are planning to give up your full-time job for a self-sufficient lifestyle. If you want to buy land and property before you start homesteading, you don't have to use all of your savings to purchase it. You have to consider other vital aspects such as renovations and improvements to be made to the property. If you are giving up your job for the homesteading lifestyle, you will have to come up with creative ideas that can help you generate money from your crops and livestock. At the barest minimum, you may have to pay property taxes and other possible utilities that will factor in the process of tending to your homestead. It also helps to put some savings aside in case of any emergency. More importantly, it is smart to try multiple homesteading income streams if you aren't just doing homesteading as a hobby or something to provide food for you and your loved ones. It is quite common among homesteaders to have up to 5 or more income streams. Some of the popular ones include selling extra produce, dairy products, wool, and other things that you can quickly get from your garden and farm animals.

When you do start your homestead, make sure you start small. Starting with something you are sure you can handle is much better than becoming overwhelmed with a backyard farm that you can't manage yourself. You don't have to wait until you have your ideal farm to start backyard homesteading. Becoming a homesteader depends more on your mindset and readiness to make homesteading a lifestyle, as opposed to the size of your land or property. If you have a window where the sun shines in, you can even start a small garden indoors by growing some spice or herbs. If your backyard is large enough, you can start a fruit and vegetable garden and begin producing the leafy greens you like to consume. Homesteading, as I have stated, is all about defining your priorities. What will be the most beneficial to you right now – a vegetable garden or a chicken coop? It is entirely up to you to decide. Some people may want to focus on chickens and other livestock for meat and egg production, while you can choose to start small with a fruit garden.

Homesteading, as you will come to realize, promotes a minimalist lifestyle. So, as you become kore and more accustomed to backyard homesteading, ensure that you continue to simplify your lifestyle. An excellent way to do this is to cut down on the incessant and seemingly uncontrollable need for the newest gadgets, mobile phones, and other parasitic things that will eat money out of your bank account faster than a bug. The whole concept of homesteading revolves around the idea that less is more. Whether many people realize it or not, there is always a cheaper and better way to get something done. Therefore, as you progress in your backyard homesteading journey, you should regularly do an audit of your life. Evaluate ways to reduce or possibly eliminate the things that are draining your money, energy, and time, so that you can reinvest these valuable resources into your self-sufficient lifestyle.

As part of your homesteading journey, you will need to learn how to preserve and store food so that they don't spoil or waste. Food preservations skills such as canning, freezing, pickling, dehydrating, smoking, and cold storage are things you must acquire to become a genuinely self-sufficient homesteader. If you follow the guidelines in this book right, you are more than likely to have more food than you require at the end of each farming season. If you don't learn how to preserve the food you produce, most will end up going to waste. You need to learn how to keep your produce from spoiling if you want to keep food on your table during the winter months. Crops are much harder to grow during winter.

Making friends with people who already have valuable experience in homesteading can help you along your journey. Although backyard homesteaders are often incorrectly misjudged as antisocial people, the truth is that they are friendly people who are eager to trade the knowledge they have with other people. Having a like mind who is also interested in homesteading can really help in case of any concern or question you have along the way. Experienced homesteaders know the weather, laws, planting and growing conditions, and other information that you will find extremely beneficial. Furthermore, a homesteading buddy is all the emotional support you need because many different people will undoubtedly think you strange. Networking is vital in homesteading.

Before you dive straight into building your fruit and vegetable garden, you have to design your homestead in preparation. So, how do you go about that? The next chapter focuses on answering this question.

Chapter Three

Designing and Building
your Backyard Homestead

Congratulations! You have successfully created a plan and set achievable homesteading goals. You now know everything you need to before you get started with the homesteading journey. Now, the next big step is to plan, design, and implement the ideal farm for your backyard space.

Brainstorming and coming up with design ideas for your new urban homestead is a fun and challenging experience. It is sometimes difficult, but that is only when you don't know the right steps to take. Whether you are a complete beginner to homesteading or you already have some form of experience with gardening and animals, there are still some basic things that you have to figure out. Getting carried away from your defined goals is super easy when you just begin backyard homesteading.

When you have a plan in place to lead a self-sufficient life, you might consider jumping right in without considering everything involved. Resist that urge as it is destructive. You can't jump straight into homesteading unless you build the perfect homestead first. Now, "perfect" in this context doesn't outrightly mean perfect. Instead, it means making a homestead that is just right for you and the backyard space you have. With all the excitement of starting their own backyard farm, many people don't consider the structural layout of the said farm. Often, things don't turn out so well as they soon come to realize that their gardens are not ideally suited, with a lack of sun, insufficient space for what they want to do, and other mistakes.

After you set the goals for your homesteading venture, you can start designing your homestead. But first, there are some essential steps you must take. Skipping these steps and jumping right to design and implementation will lead to regrets in the future, or failure, in the worst scenario. The first crucial step, which I briefly talked on in the previous chapter, is to check out the zoning laws for gardening and animals in your region. In most states, there is a required distance between neighbors and regulations in place for the number of animals you can keep in your home, and possibly the type of animals too. Many states don't allow residents to grow gardens in their front yards also. To find out the zoning laws in your state or location, Google the provincial laws of that state. If Google doesn't provide the information you need, go to the local registry to find out.

A vital step to take before you start designing or even building your homestead is to make sure that the garden sunlight exposure is just right in your backyard. For most of your vegetables and fruits, you will need more than 8 hours of sunlight. The required sunlight hours vary from plant to plant. Some vegetables don't need more than 4 to 6 hours of daylight every day. Generally, however, 8 hours is just right for most plants. Choose one day to check out the sunlight and see the outbuildings that are likely to cast shadows. If you are planning to raise some chickens, for instance, you have to build a chicken coop. Part of the design session is to position your chicken coop in a place where it won't cast shadows over your fruit or vegetable garden, thereby preventing the plants from getting the much-needed sunlight.

Since you will be raising livestock, you have to factor in their needs before you begin designing or building your homestead. Do you need to make some buildings for the farm animals you want to raise? What kind of animals do you even want to raise? Do you need to build an outbuilding where you can keep things such as hay or milk the goats (if any)? The size of the buildings is a crucial thing that must factor in your brainstorming sessions. You might decide to start with just six chickens, but later on, you will undoubtedly choose to expand your flock, which means you will need a bigger chicken coop than you have for 6 chickens. Fencing is also essential. A large and tall fence is vital for keeping pests and predators out of your backyard homestead. An urban resident may need rodent-proof garden beds. Plus, most backyard homesteads need large and tall fencing to keep the livestock and the garden safe. All of this will be quite costly.

Another thing about raising animals is that you have to think of their compatibility when designing the backyard farm. Usually, goats and sheep are great companions. Still, you must ensure that breeding males can't access the females at any time. And you definitely shouldn't allow them to cross-breed. This is very important. Chickens, on the other hand, can be let free around sheep, goats, and cows, but you need to give them access to a lot of areas in the homestead because they like to free-range. If you keep your chickens in the same place as your milking animals, their poop may get in or around an udder, which could be really bad. If you are keeping the milking animals for their milk, you need to be careful because there is an increased risk of E. Coli and Campylobacter. This doesn't mean that chickens can't hand out with animals such as goats and cows, though; it merely means you shouldn't keep them locked in the same building. Chickens should have their own coop where they sleep and lay eggs. Also, if you want to keep pigs, know that they are omnivores animals. This means they will likely hunt down and consume other animals on your farm if you provide an opening for them to do that. As a result, you need to make sure that the pigs are with their own kind.

Animal containment is another factor to put into consideration. Since meat chickens become fully grown once they are 10 to 12 weeks old, you have to start by raising chicks, i.e., the young chickens. So as not to restrict them to a small area that can be quickly run down by all the feet of your other farm animals, it's better to get or build them a chicken tractor till they are older and more mature. A chicken tractor is a moveable cage that lets you move your chicks to new grass spots each day. Many people believe that meat rabbits should be put in small cages, but it much better for the rabbits if you raise them in their natural habitat, which is the grass. So, you need to construct a rabbit tractor – similar to your chicken tractor, but with chicken wire. The chicken wire will help make sure that the rabbits don't dig a way out of the tractor. When designing your homestead, you also have to think of the environment. You should make your own compost when you start backyard homesteading. If you want to do this, the best thing is to put the compost pile near the animal buildings. It is even much better when you have your garden close to the animal pens and the compost pile. That makes things more convenient for you.

You also have to think about homestead trees in the backyard area. Planting bushes and trees around your backyard farm can be costly, but you will find it worthwhile in the long run. When you plant your trees and shrubs, they have to be in spots where they can't cast shadows over other areas. Thankfully, many fruit trees have semi-dwarf and dwarf sizes, which means they won't take that much space in the backyard farm. Permaculture design can also help integrate your gardening and animal husbandry experience. Basically, using this design means you want your animals to get some of the jobs done for you. For instance, chickens can help with composting, and they can clean up the fruits in your orchard when you need them to. Goats can help clear out new areas for you to plant crops on later. With the right knowledge, you can make permaculture work on your backyard homestead.

Obviously, there are many factors to put into consideration before you even get started with the planning and designing of your new backyard homestead. Once you have considered all the possible restrictions in terms of land and budget, you can go ahead to design your homestead. Whether you have a large acreage for the new farm or a small urban lot, you have to map out the land. After considering your goals, you can proceed to design multiple different combinations for their farm.

Don't do the backyard farm design all by yourself. You either do the brainstorming sessions with your family, or you keep them in mind when you are brainstorming for the perfect homestead idea. When conceiving ideas, don't focus on the place you are in life right now – think of where you want to be in the next ten or more years. Do you have a family? Or, are you single? Do you want to have more kids in the future? Or, are you going to be empty nesters? All of these are crucial information because the design creates must accommodate your lifestyle and needs. Even if you think there are little chances of your family situation changing in the future, you should know that specific changes are inevitable. Not only this but if you want to design a homestead farm that can survive for decades and more, you need to design a space that is both functional and appealing.

Start by drawing an overhead layout and try your hands at different ideas before you settle for one. You may need to doodle and design a lot of combinations before you finally get a design that you think is right for you. This is totally normal. Most homesteaders create a lot of sketches before they finally settle for one. Your plan will change depending on your budget and time. To make your job easier, you can try a garden planner app to get the foot pages and every detail you need in the garden plan. Whether you live in a condo or you have a good-spaced backyard that can be put to good use, a garden planner will help you determine the perfect size and structure for your backyard farm. The great thing about these garden planners is that they usually give you the layout and design in a document you can print. So, once you design your backyard farm, you can print it out and get to work.

Backyard homesteading is not half as difficult as many people think it is. Still, it does require lots of commitment and determination to succeed. In the second part of the book, we explore everything there is to know about starting a successful vegetable garden that can provide you all the vegetables you and your family need.

PART II

Vegetable Gardening: Everything You Need to Know

Contrary to what you believe, growing vegetables isn't tricky, but I'll admit it is a little complex. Unless you know what to do, you can't just find your way around planning and starting a vegetable garden. The problem with most people who want to start a vegetable garden is that some are looking for a quick solution – something that can help them make the perfect vegetable garden plan, which will make their vegetable garden yield as much as they want. Others don't mind investing the amount of time needed to build the garden. Still, they are puzzled by the plethora of layout designs and plant combinations.

When planning a vegetable garden, it is super easy to dive right in and grow as much as you can in a single season. Ask any experienced homesteader or gardener, and they will tell you that you are setting yourself up for disappointment when you do this. Growing too many vegetables than you can handle as a newbie will give you an overwhelming amount of weed to maintain. You are very unlikely to handle that well in your first year. The best thing to do when planning your vegetable garden is to make a list of all your favorite vegetables, specifically those that you consume regularly. Once you make a list, you can narrow it down to the ones that are most expensive to purchase in the grocery store, or those that you undoubtedly enjoy the most. In your plan, ensure to note that you will be creating new vegetable beds every year. This means that you should be expanding the better and more confident you feel in your vegetable gardening abilities and skills. Also, research the best timesaving shortcuts that are perfect for you and your lifestyle.

If you have never done anything regarding vegetable gardening or even own a garden at all, you will likely use a completely new area for your new garden. So, you also have to decide on the planting and garden system you will use for your new garden. Thankfully, there are different types of planting system that you can use for your modern garden. These include raised beds, square foot gardening, traditional rows, etc. As an urban homesteader, you want to make sure that your planting system makes use of the available space most attractively and sufficiently. At the same time, you must make sure that your plants get all the exposure they need to grow healthily. To ensure this, consider the size of your backyard space. If you have a large space and you want to grow as many vegetables as possible, you might want to use the conventional planting system. But if you have a medium or pocket-sized space available for gardening, use any of the other methods. You can also use other methods if you are trying to grow vegetables in the most inexpensive way possible.

When you just move to a new home, you may not pay attention to the health and condition of the soil there, but this will always be an important factor when you start growing fruits and vegetables. Poor drainage will make it difficult for the garden to thrive, and heavy or thin, stony soil is usually difficult to grow plants on. The best way to improve the condition of the soil is to dig deep and thoroughly and bury a large amount of organic matter into the soil. But the most preferred planting techniques for urban farmers is the use of raised beds. Your planting system will also be determined by the types of plants you want to grow. Raised beds are usually great for planting vegetables. Quantity and spacing should also be considered when choosing a planting system. Obviously, you need to produce enough vegetables to last your family for as long as you want. But you must also pay attention to the instructions on your seed packets to ensure that you leave enough space between your vegetables so that they can grow healthily.

Finally, crop rotation is an important practice in subsistent farming. No matter the gardening system you choose, you need to make sure you don't plant your vegetables in the same spot every year. This is to ensure the health of your vegetable crops. Rotating your planting spaces each year obstructs the development and build-up of crop diseases and pests, and also allows the soil to replenish the nutrients needed by your crops. Normally, these groups of crops are rotated together each planting season.

- Leafy plants: Lettuce, broccoli, spinach, cabbage, cauliflower.

- Root plants: Garlic, onion, turnips, potatoes, carrots, beets, radishes, shallots.

- Fruiting plants: Tomatoes, eggplants, cucumbers, zucchini, pumpkin, sweet corn.

However, this approach is simplified. It ignores the fact that some of the vegetables, however dissimilar, belong to the same family and may suffer from the same diseases. For instance, tomatoes and potatoes belong to the same crop family, and they can both be affected by potato blight.

What are the planting systems that you can use for your vegetable garden?

The traditional vegetable garden is the most commonly known gardening system. It involves growing your vegetables or crops generally on a large patch of soil. This system still works for many people, but it isn't the best option for an urban farmer, especially one who has to make use of a smaller space. But if you have the time and space, you can use the traditional gardening system. To use this method, you have to clear your land of all weeds and dig thoroughly before you add as much organic matter – specifically compost and leaf mold – as possible. It also helps to add paths through the middle of the plot, and these paths should be vast enough for a wheelbarrow to go through.

Permaculture is another gardening system that is all about sustainably using your land while harmonizing with nature. The permaculture technique is based on the concept of "Reduce. Reuse. Recycle." Using this gardening system helps reduce your carbon footprint on your farm. The whole idea of permaculture gardening is to build your garden in an easily accessible way that minimizes the need for labor and to grow your crops without the use of chemicals. You have to check out the sunny spots in your backyard, as well as the sheltered areas and the direction of the strong wind before you decide on which growing method will be suitable for those conditions. Permaculture gardening involves reducing wastage and using systems like compost bins, water butts, and wormeries. You can easily integrate permaculture practices into your vegetable garden, no matter how big or small, while following any of the techniques mentioned above. In a subsequent chapter, you will get guidelines on how you can use each planting methods.

No-dig is a planting system that was developed by people who think digging is difficult work for obvious reasons. But that isn't the only reason why no-dig was developed. People also felt that digging causes light soils to lose moisture quite rapidly while distributing weed seeds. However, the no-dig gardening technique isn't recommended for you if your soil is highly compacted. To succeed with this method, you have to create narrow garden beds that are at least 15cm tall between boards. Also, the beds must be held intact with pegs that are buried deep in the ground. You also need to spread out multiple layers of newspapers over the soil, with a mulch of sawdust, straw, and grass clippings included. Make sure you thoroughly water the bedding before you add one layer of compost and then round it off with around 6cm of soil, which is where you need to plant your vegetable seeds. Naturally, the soil level will lower as the mulch layer rots, but you can always add more compost to top up the beds, as needed.

Raised beds follow the same principles as the no-dig technique, but is much deeper. The beds in this technique are big boxes of soil mixed with compost. You can make the beds from materials such as wooden crates, boards, bricks, or railway sleepers. Then, you fill the raised beds with compost in a way that it is higher than the surrounding soil to make it remain as dry as possible. The raised beds technique prevents the usual problem of poor drainage and bad soil. While most of the garden will be taken up in paths between the raised beds, it allows for easy access to the vegetables and prevents the soil from becoming compacted. The deeper soil acts as compensation for the lost space. If you are very deliberate in the planning phase, you can easily incorporate certain systems to cover the beds with cloches and provide warmth during the cold season. You can purchase some of the commercially-available systems which come with holes for hoops and allow you to cover your whole bed with netting.

Square foot gardening is the final planting system that you can use for your vegetable garden. This system is perfect and effective for you if you have landed at a premium. The square foot gardening system involves dividing a specially built raised bed into several one-foot modules, planting each of your vegetables into that particular area. This method is suited for salad vegetables and other miniature vegetable varieties. The close planting system produces a climate and environment that suppressed weed growth. The crops are easily accessible from all sides, making it easy to grow your vegetables directly outside the kitchen door.

The best thing about vegetable gardening is that you can combine two or more of these methods if you think that you need a system that suits you just right.

Which planting system is best for backyard homesteading?

Everyone knows that building up your soil is the singular most important factor in vegetable gardening. An organically-rich and deep soil quickens the growth of healthy and long roots that can reach more water and nutrients. This, in turn, results in the extra-productive growth of your vegetables. The quickest way to get the deep layer of a rich soul that you need is by making raised beds for planting. Raised beds often yield more than double the amount of space than you get when you use the traditional planting method. This is due to the loose, fertile soil, as well as the effective spacing. Using less land for paths gives you extra room for growing the plants.

Also, raised beds to save more time. According to research conducted, you only need about three days of work to plant and care for a 30-by-30-foot garden using the raised bed techniques. In this timeframe, the researcher was able to harvest around 1, 900 pounds of vegetables, which is nearly a year's worth of vegetable supply for a family of three, from the garden. The one reason why raised beds can save more time and produce more crops is that there is very little space for weeds to grow between the vegetables – meaning you don't have to spend the bulk of your time weeding. The effective spacing also allows you to water and harvest more efficiently. The shape of the garden beds makes a big difference, as well. Raised beds are more efficient for planting when you round out the soil to create an arch shape.

Companion Planting

There are varying crop layouts that you can use for your vegetable garden. And when harvest comes, there will be far more varies as a result of factors that are outside of your control. Companion planting is all about mixing up your vegetable plants to disorient pests. Pests are usually attracted to a large patch of one crop, whereas mixing up the plants confuses them. The only time you shouldn't use companion planting is when you are growing plants that require special protection, such as broccoli, cauliflower, cabbage, etc.

Vegetable gardening isn't hard, as long as you know and follow the guiding principles. To succeed, just keep in mind that overcrowding your vegetables or planting them in low-quality soil is a recipe for failure. Also, the principle of crop rotation is one that you should take very seriously as it can help you learn constraint.

Chapter Four

Best Types of Vegetables to Grow in Your Home Garden

Your up-and-coming vegetable garden should be a dependable and abundant source of your essential nutrients. But it isn't just about the nutrients; it is also about the rate at which your vegetable plants grow. After all, what is the point of starting a backyard vegetable garden if the vegetables aren't growing fast enough for your consumption? Starting a vegetable garden isn't just about growing any, and all types of vegetables that you think should go in a garden. It is also about making sure that these vegetables are foods that you genuinely enjoy. You should plant your favorite vegetables. But you should also make sure that your vegetable selection contains mainly fast-growing greens that can go on your table as fast as you need them. As a newbie to vegetable gardening, making sure that the vegetables are also easy to grow and harvest is another important thing.

Before you plant a vegetable garden, the biggest question you will undoubtedly ask yourself is, "What types of vegetables should I plant?" Although it largely depends on what you want to plant, ensuring that you begin gardening with veggies that are really easy to grow is key to your homesteading success. Well, there are two types of vegetables: cool-season and warm-season. Knowing the difference between the two types of vegetables is essential as it lets you know the right time to plant each type of vegetable on your planting list. Although expert gardeners might think that everyone should know the difference between cool-season vegetables and warm-season vegetables, the truth is that most beginners usually don't know and that is one of the reasons why they fail with vegetable gardening. At the most basic level, the difference between cool-season and warm-season vegetables is the time of the year you plant or grow them.

Cool-season veggies are those vegetables that grow during colder seasons like winter, fall, and spring. Warm-season, on the other hand, are vegetables that love the heat of the summer sun. But there is a lot more to cool-season and warm-season vegetables than the time you grow them. As you can tell from the name, cool-season vegetables love the cooler climate. The general rule of thumb is always to plant your cool-season veggies to mature before the day time temperature reaches or goes beyond 80-degree Fahrenheit. The ideal temperature for growing cool-season vegetables is between 55 and 75-degree Fahrenheit. Plus, cool-season veggies require lower soil temperature to germinate. The seeds start to germinate with a temp that is as low as 40-degrees, but the highest is usually around 70-degrees. Cool-season vegetables can fare well in much cooler temps, and they can even survive a bit of frosting without any damage. In fact, some cool-season veggies like Kale and Brussels sprouts actually require a little frost to improve their taste and quality. Hot weather is bad for cool-season vegetables. Any temperature that exceeds 80-degrees will adversely affect the health and quality of your cool-season vegetables. Hot weather makes the leaves of some cool-season vegetables bitter while causing others to bolt, i.e., set seeds and send up flowers. Therefore, you must always plan your vegetable accordingly. Your cool-season vegetables should be planted early around springs so that they can fully mature before the summer heat descends. Planting times will vary based on the hardiness zone your location is. Keep in mind that you can also plant your cool-season veggies during the fall.

Examples of cool-season vegetables include:

- Beets

- Lettuce

- Turnips

- Carrots

- Broccoli

- Cabbage

- Leeks

- Spinach

- Kale

- Cauliflower

- Green onions

- Radish

- Leafy greens generally

Warm-season vegetables are the summer greens. They grow best in hot weather. Plant warm-season vegetables in fall or springs, and you will find that the weather isn't great for their growth. Cold weather or frost usually damage the health of warm-season vegetables, or in worst cases, kill them. Due to this, you shouldn't plant your warm-season veggies until the day time temperature is between 65 and 95-degrees. Besides, warm-season vegetables require a soil temperature of at least 65-degree Fahrenheit before they can germinate. But you should know that they also don't mind a soil temperature that is much higher than that. The planting times for warm-season vegetables start from mid to late spring. While some cool-season veggies can handle a little warmth, warm-season veggies cannot tolerate below 12 hours of daylight. So, your warm-season vegetables have to be planted in the sunniest parts of your garden. You should know that you will face more pests and watering problems with warm-season vegetables than cool-season ones. So, ensure that you build a solid drip irrigation system to provide your warmth-loving veggies all the water they need.

Examples of warm-season vegetables include:

- Pepper

- Cucumbers

- Onions

- Pumpkins

- Mustard

- Potatoes

- Tomatoes

- Summer squash

- Squash

- Melons

- Sweet potatoes

- Eggplant

- Herbs

One more thing you should know is that you can "modify" a little with warm-season vegetables. There are great ways you can get an early start on growing them, including the use of hoop houses, cloches, wall-o-waters, and cold frames. For instance, using wall-o-waters allows you to plant your tomatoes two months before the last frost in your region. Doing this can increase your harvest to a great extent. Also, using a heavy fabric cover to protect your warm-season vegetables in colder temperatures can add an extra one month to the harvest of vegetables such as tomatoes and other warm-season veggies. Some crops have more hardiness than others and some herbs handle cool temperatures really well. Broccoli tolerates heat better than most cool-season vegetables. So, you need to have some flexibility with your planting schedule.

Having said that, below are some fast-growing and easily-grown vegetables that are great for you to start your backyard homesteading journey.

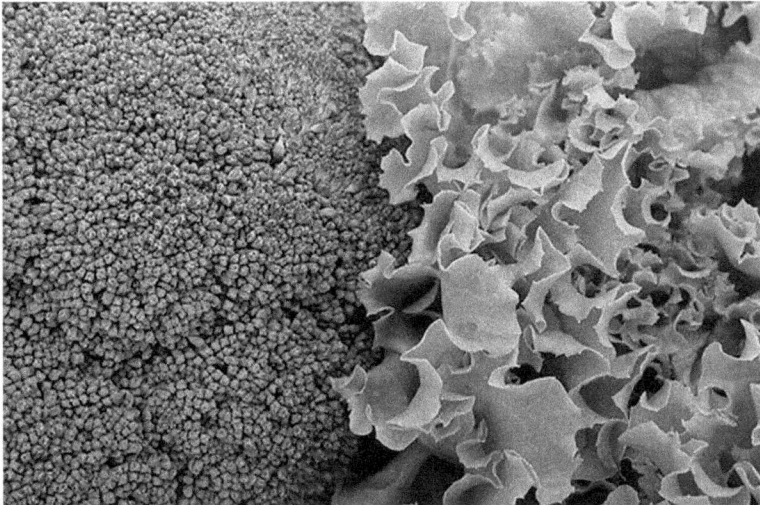

Arugula

If you want some bold-tasting leafy green, arugula is the right choice. Plus, it a fast-growing vegetable that can be harvested in a way that maximizes its output. The best time to plant arugula is a couple of weeks before the last frost. If you see flowers growing on your arugula plants, it means that the end of the growing season is near, and the harvest season is coming. Make sure you remove the flowers from the plant to help them produce some more. The best time to harvest for young leaves is 20 days after planting, while you should harvest 40 days for mature leaves.

Spinach

This is a highly nutritious and delicious vegetable that should be part of any backyard garden. Spinach is best planted during the cooler weather; this helps you to maximize the output of the plant crops. Although it may fare fairly in early summer, spinach cannot survive hot weather. You should plant the first spinach crop a month to your expected frost date, and the second crop should be planted 6 to 8 weeks after the last frost date. Spinach can be harvested as early as 20 days after planting, or as soon as the leaves are suitable sizes for consumption. But make sure you get rid of the outer leaves first and let your harvest extend into the season.

Kale

Kale is probably the easiest vegetable to grow, plus it is incredibly nutritious. Expectedly, Kale grows better in cooler temperatures like many leafy greens. In fact, they can survive down to 20-degrees Fahrenheit. Additionally, Kale becomes sweeter, the more the weather becomes chiller. The best time to plant Kale is in the spring just one month before the last freeze date, and the second crop in early fall or late summer. Kale should be harvested when it is 60 days to maturity, while the young kale leaves should be harvested once they are the size of your hands. Ensure you leave 4 to 6 leaves on the plant if you harvest while it is still maturing. In the USDA zones, Kale can be grown through the winter months.

Radish

Among the easiest vegetables to grow is radish, and the good thing is it comes in spring and winter varieties. The spring radish typically grows faster than the winter variety. To monitor their development, gently pull on the leaves or push the soil aside to check if bulbs are developing on the plants. Fortunately, radishes multiply in incredible quantities. Although their nutritional level is just on the normal range, they actually grow incredible crop sizes. Radishes are best planted about a month before the last frost date. To harvest them, you have to wait for 20 to 30 days for the spring radish and about 60 days for the winter radish.

Lettuce

Everyone knows lettuces are a delightful addition to any diet. They are cool-season veggies, and they respond very well to every selection harvesting method that is used on other leafy greens. Lettuces are best planted in the spring and fall seasons for the best results. The spring crops should be planted a month before the final freeze date, and the fall crops should be planted a month before the first frost date. Harvest time is usually around 40 days to maturity, but you can harvest lettuce leaves at any point of growth.

Green Onions

If you need something to add flavor and spice to your meals, then you need to add some green onions to your vegetable garden. But that's not all – they also have great nutritional advantages for you. Adding some green onions in your soup can make the sauce sing, plus you can add the stalks to your meat dish. The only limitation in adding green onions to your meals is your imagination. Onions are generally warm-season vegetables; so, it is best to wait until some periods after the last frost date to plant them. Onions can be harvested around 50 days to full maturity. You may also harvest the stalks as soon as the leaves are at least 5" high. However, doing this may affect the bulb's growth.

Carrots

Homesteaders know that carrots are somewhat tricky to have in a vegetable garden; this is because their growth happens out of your sight. The trick to growing carrots the right way is to plan them the right way. For starters, you need to grow the carrots in loose, sandy soil, and be careful not to apply too much fertilizer in the early stage of planting. Ensure that you plant enough carrot crops so that you can monitor their development with no fear of losing the crops. Carrots are best planted about one month before the last frost. Also, you shouldn't harvest them until they have had 60 days to mature. Once your carrots are around half of an inch, it means that they are ready to be harvested.

Turnips

Turnips are a great source of antioxidants and fiber. They have been around for thousands of years, and are known to be easily-grown, nutritious, and reliable in any garden you find them. Some people even use turnips as alternatives for potatoes. For the spring crop, you should plant turnips around 3 weeks before the final freeze, and then you can plant some in September for fall crops. Warmer regions in the U.S. can plant and grow turnips throughout the winter months. You can harvest your turnips as early as thirty days after planting. Still, they usually don't attain full maturity until they are about 60 days old.

Tomatoes

Tomatoes are many homesteaders' favorite vegetable to have in their gardens. To get the best results on growing, harvesting, and collection of seeds, you should buy some indeterminate heirloom tomato seeds. Heirloom tomatoes are known to produce very delicious crops, and they also take from seeds easily. Indeterminate tomatoes will continue to grow until they are killed off by the cold climate. Tomatoes should be planted after the final freeze in the spring season. Start planting your tomatoes 2 to 4 weeks before the final freeze indoors, if you can, as this is a great technique for achieving early production. Tomatoes are usually harvested after 60 days.

Pepper

Whether you grow hot peppers or bell peppers, just make sure you have some pepper in your backyard garden; they always come in handy. You may even decide to mix the peppers you plant, resulting in cross-pollination, which will give you some sweet peppers with a little heat in them. Fortunately, there are lots of recipes you can use your peppers for.

Cucumber

Cucumbers can be harvested about 12 to 14 weeks after you plant them. Normally, you should try to harvest them early in the morning when the weather is still out cool. Also, cucumbers should be harvested at a young age because they tend to turn bitter as soon as they start to show symptoms of seed production.

Fast-growing and nutritious vegetables are some of the best ingredients to include in your backyard homesteading plan. Once you become more accustomed to vegetable gardening, you should be able to grow more difficult vegetables that you need in your home. A reliable source of readily-available vegetables for your dining table is something missing in many homes. Still, it no longer has to be missing in yours. Whether your goal is to reduce your grocery bill, spend more time with nature, or live a more sustainable and self-sufficient life, there is always an excellent reason to start your own vegetable garden. With so many tips to make vegetable gardening an easy and relatively stress-free journey for you, you really have no excuse as to why you can't start growing your own food in your own backyard.

Now that you know the best types of vegetables to include in your backyard garden, let's get to the actual planting, which is a whole lot of fun than the planning and preparation phase. It's time to go into the backyard and get your hands dirty so that you can relax and enjoy the fruits of your labor. Literally!

Chapter Five

Planning and Planting Your Vegetables

Planting vegetables is easy when you already have the knowledge that you have gained so far in this book, but it gets even easier. The key to building a successful vegetable garden is to find the perfect location. As stated in a previous chapter, you have to consider the sunlight exposure, soil, drainage, and access to a good water source. Where you grow, your vegetables will depend largely on the types of vegetables you want to plant. As you have learned, some vegetables require more sunlight exposure than others. Some fare better in the shade. Most garden vegetables fall in between the need for sunlight and shade. You need to consider the extent to which you want your vegetable garden to affect the overall aesthetics of your backyard garden.

Regardless of the planting technique, you want to use for your garden, the best place to start is to select a portion of your land where you want your vegetable garden to be located. Then, you can mark that space with spray paint, a flag, a piece of board, or anything that can remind you of where the edge of the garden should be. Then, you can use the tips that have been given in previous chapters to determine if that spot is perfect for your garden. Once you determine this, as well as the space available for planting, the next big step is to prepare your soil for planting.

If your land is blessed with sandy loam soil with a few rocks here and there, preparing your soil should be easier than you think. If not, which is the likeliest option in your case as an urban resident, preparing your soil might be the most challenging and hardest part of vegetable gardening that you have to deal with. First, turn over your soul. Unless you have natural loamy or sandy soil or your garden will be on a spot where there was once another garden, you have to check out your soil and get rid of the large rocks that must be buried deep in the land. If you own a tractor, you can easily do that on your own. If not, you just have to find people who are willing to rent out their tractors and other implements. Some people are quite okay with doing homestead chores for backyard homesteaders as a regular source of income. If you can't find anyone, you can ask around your local area; at least one person must know someone that plows gardens for a couple of dollar bills. If you maintain your garden well enough, that will be the only time you will ever need to plow it.

Rock picking is part of the planning and preparation phase before you actually start planting. This doesn't require much explanation, because it is exactly what you think it is. Find a wheelbarrow or a large bucket and start picking your way through the ground that has just been plowed. Pick any stone that is more than an inch in size. If you have enough energy for it, you can pick up rocks that are much smaller than that. However, stones below one inch generally won't affect your roots or damage your tillers when you start growing your vegetable garden.

To take it the scientific way, you should have a sample of your soil tested to determine the pH level. If the soil in your backyard is of the clay type, you will need to mix in some sand, organic matter to make it suitable for gardening. But don't mix in chicken manure if you want to start gardening right away because this can burn your tender vegetable plants. Since your garden will probably only be a few square feet, you can make use of a hoe or any other gardening hand tools. However, if you want to expand your garden to become larger, you may want to invest in a brand-new tiller or an already-used one.

Once you are sure that your soil is of good consistency – grab some handful and roll into a ball. When you open your hand, the ball should feel loosely packed and be easy to break up. Then, you can reach out to the Farm Services agency in your area to get a soil test done. The specialist there will give you information on how you can collect a suitable soil sample to submit for evaluation. Then, they will offer you instructions on how you can optimize your soil. You may be told to add some lime or sulfur. Even when you have worked really hard to improve the soil in your backyard space, you may be disappointed to find that your soil may still not be composited enough for your vegetables. For example, you may have too much clay in the soil mix, which will affect the vegetables you can grow. You may be able to grow an abundant amount of peppers, tomatoes, and other vegetables that grow above the ground level. But you will find it hard to grow vegetables such as onions, potatoes, and carrot. Fortunately, the more you use your vegetable garden, the better the soil will become over time. The best farming practice is to grow whatever is to plant the vegetables that grow well on your land. The other thing you can do is to haul in a considerate amount of sandy loam topsoil to improve the condition and health of your soil.

Most vegetables are annual crops. They last a season and then you have to plant them again in the following year. Some vegetables, such as onions and tomatoes, may even pop up in strange areas in the following years if you leave some yield to rot in the garden, and their seeds are scattered everywhere. But usually, new plants have to be planted each spring. Although we have discussed the fastest-growing vegetables that you can grow in your new garden, don't forget that it all boils down to the type of vegetables that you like to consume. Also, remember to research the vegetables listed above to ascertain whether they suit the climate in your local area or not. Once you are sure that the vegetables you want are suitable for planting in your climate zone, you can go ahead to purchase the plants or seed and begin planting. To determine the best vegetable varieties to have in your garden, you should check the seed packets. But, as I have certainly mentioned, heirloom vegetables are often the best varieties to plant – especially as a beginner to backyard gardening.

Step-by-Step Planning and Planting Guidelines
To place plants in your new vegetable garden after preparing the soil, follow the steps below:

- **Tender plants:** The tender plants are those vegetables that are the fussiest. Some of them include peppers, tomatoes, eggplant, basil, etc. Unless you have an extremely warm or hot climate, the sunniest spots in your garden should be reserved for the high-value plants, so they should be the first on your list. Walls that face the southside are particularly great for providing the sunlight and heat required by the plants to multiply their harvest.

- **Roaming plants:** Next, you should have the plants that like to spread vines around the garden. These include vegetables like squash and melon. The roaming plants should be located at the edge of the garden beds so that the broad leaves around the vines don't cover up other plants that need sunlight exposure in the garden. Planting them at the edge of your vegetable bed allows them to spread out across the grass and paths.

- **Vertically climbing plants:** Any vegetable that grows up to support other plants such as beans, peas, and even squash should be located in the part of the garden where they can't shade other plants. An example of a vertically-climbing vegetable is the cucumber plant. But there is an exception during very hot summers when you need to put your cool-season veggies such as spinach and lettuce can hide under the shade when the weather is at the highest during the day.

- **Irrigation:** Some plants perform really poorly when the condition is dry. They include onion, celery, etc. The parts of the garden that are slightly lower than others are capable of retaining more water and moisture. Otherwise, you will need to find a way of providing irrigation for the vegetables that really need it. This is to ensure and promote consistent growth and health.

- **Pollination:** Some vegetable plants need to be planted near others to pollinate as they should and produce you the edible produce. An example is sweet corn, which needs to be grown in blocks for it to produce full corn cobs.

- **Accessibility:** Which of your vegetables do you want to harvest regularly? Tomatoes, salad, vegetables, or herbs? These should be planted near your kitchen for easy accessibility. With easy access, you will not only be able to use them as you want, but it will also be easier for you to keep weed growth in check and get rid of slugs regularly.

- **Succession planting:** If you don't have enough space, but you want a particular crop throughout each season, you can use intercropping or succession planting to get all the vegetables that you need.

- **Overcrowding:** As tempting as the idea may be, you should never grow more plants than your garden can contain. Overcrowding is easily the number one mistake made by many new gardeners. This is understandable since vegetable plants are usually very small as seedlings and gardeners generally detest the idea of pulling up the product of their hard work just to thin them out.

In simplified terms, here are the ten basic steps to planting your vegetable garden.

1. Choose the perfect location with enough space, sunlight exposure, and proximity to your water source. Make sure the garden area is leveled to prevent erosion.

2. Select the veggies you want to grow, based on your space, climate, expertise level, and personal preference. Grow some of the fastest-growing vegetables such as lettuce, pepper, cucumber, carrots, and beans.

3. Prepare your soil for planting by mixing the compost and naturally-procured fertilizer. You can test the soil acidity or purchase readily-available soil from the stores in your local area.

4. Check the planting dates for the vegetables you want to grow or follow the planting dates for some of the vegetables in the preceding chapter. Note that planting information is also available on the seed packets. Review the required planting conditions for each of the vegetables you want to plant.

5. Plant the vegetable seeds in the already-prepared soil. Don't forget that you can use any of the planting techniques that have been discussed. Follow the spacing and depth directions on the seed packets very carefully.

6. Gently spray the new plants with enough water to keep the soil moist throughout the planting and growing season. You should add a spray nozzle to your hose to make the water rain-like when spraying.

7. Mulch regularly to keep the weeds out of your garden. Add at least 2-inches thick layer of mulch to the garden to prevent the weeds from overtaking the garden and killing all your crops. When weeds do appear in the garden, yank them sharply from their stem to extract the whole root.

8. Again, make sure you follow the spacing guide on the seed packets and be quick to get rid of all crowded seedlings as soon as possible.

9. Fertilize your garden as frequently as required. Till the soil gently to mix in the fertilizer and keep the soil rich. You can make your own organic fertilizer by mixing some eggshells, fish tank water, Epsom salt, and kitchen compost together. Or, you can purchase some ready-made garden fertilizer from your nearest store.

10. Pick your vegetables once they are mature to eat or when you need them.

Chapter Six

Maintaining Your Vegetable Garden

Vegetables require pretty much the same amount of care as ornamental plants, sometimes even less. But they certainly aren't as forgiving of neglect. Vegetable plants use a significant amount of energy to bloom and produce fruits that do not get to mature before they get harvested. A vegetable plant sets fruit so that it can produce more seeds. Still, we often end up harvesting the vegetables before the seeds become fully-formed. This can be very stressful for your vegetable plants; so, it is vital that you provide them exactly what they need and more to keep them in good health and vigor for maximum production. Neglect of your vegetable plants will often result in lower yields and poor-quality crops as a result of pest problems. The guidelines we will be discussing below will help keep your vegetables healthy as they grow along the season to give you a high-yield return on investment.

The first guideline is to water your vegetables regularly. Watering is an essential part of gardening, and the importance cannot be overstated or understated. Regular water is just as important to your plants as sunlight exposure. This means that you must give your plants at least one or more inches of water each week, and more when the weather is exceptionally hot. Without regular water, your vegetables won't fill out as they should, and some, such as tomatoes, will crack and burst open if suddenly showered with water after being denied for a while. Since you can't always expect rainfall to help you, it is better to install a drip irrigation system if you have the means. Most of the new systems on the market are easy to install, and they are very affordable. You will be able to save money on your water bill as well, because the water from your irrigation system goes directly to the vegetables' roots, making it almost impossible for you to lose water to evaporation. If you don't want to install a drip irrigation system, you can locate your garden near a reliable water source, such as a water spigot. Watering is much easier when you don't have to drag the hose out.

Just as you have learned, you have to regularly get rid of excess seedlings to ensure that your garden remains very healthy. This is especially important when the plants are being sown directly from the seeds. This process is called thinning and is an essential part of vegetable garden maintenance. Many gardeners find it difficult to sacrifice their seedlings, but leaving all the sprouted and unwanted seedlings to grow closer to your healthy plants can stunt their growth and reduce the garden yield at the end of the season. Once true leaves come out of the plants, remove the seedlings to ensure that the vegetables remain at the required spacing distance. If you can't get rid of the extra seedlings without affecting the roots of the remaining plants, simply get rid of the seedlings at the soil line. Let the strongest seedlings remain.

Staking the plants is also a maintenance task that you must regularly perform at the start of the gardening season. Tall and climbing plants such as cucumbers require you to stake or trellis them. The best thing is to do the staking during the planting time. If you don't do it until the plants have grown beyond staking, you might end up injuring the plant roots. So, the staking of the vegetable plants should be done early in the season. Later on, you will need to prune suckers if you have some tomatoes in your garden. Pruning tomato suckers involves getting rid of the growth that takes place in the place between the stem and branch. If you leave the suckers to grow, they may end up becoming other stems with flowers, branches, fruit, and other suckers, which will all be competing for the same nutrients with your original tomato plants.

Veggies don't like it when you leave them to compete with weeds for the little food and water they have. Each gardening season starts on a blank slate after you ready the garden beds. Therefore, you must keep up with the weeds as soon as you have planted your vegetables. Doing this will help ensure that your crops stay in the best shape. If you get rid of weeds as soon as they appear regularly, it will never get to the point where they are out of your control. In addition to removing weeds from your garden, you must also remove nearby weeds along the surrounding pathways and the grass around your garden. If you allow the surrounding weeds to go to seed, they may just end up taking over your garden. Keeping weeds in check right from the start of the gardening season is one way to ensure you won't need to use herbicides later in the hotter climate.

Mulching is one of the most important tasks that you must perform for your plants to stay healthy. Mulching your plants help suppress the growth of weeds, cools your plants' roots, and saves water. When the plants become dense enough, they can even serve as their own mulch. The best type of mulch for vegetable gardens is seed-free straws. It is good as a cover, and also easy to push aside when it is planting time. Plus, you can turn it into good soil once the harvesting season ends. An extra perk of mulching is that spiders love to feast on garden pests while hiding in the straws.

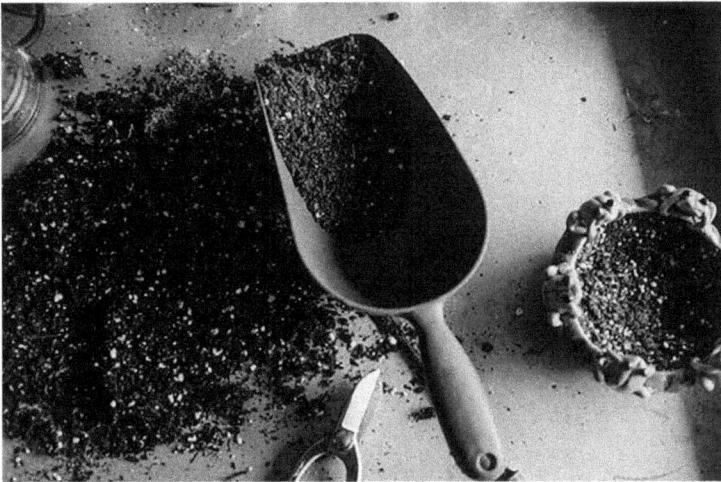

Finally, make sure you take steps to enrich the soil. Vegetables feed heavily, which is why you should mix in some organic matter into your garden every year before you plant. Furthermore, you should dress in the more organic matter once or twice while the growing season is still going strong. Of course, different plants have different needs, which means you should take note of the fertilizing guidelines that come with the seed packets or seedlings. Organic plant foods release slowly, and they can help your plants stay fed all through the season. If you choose to go for a water-soluble fertilizer, ensure that the garden is watered well before you apply it to the soil.

Since getting great soil in your garden takes so much hard work, it is only reasonable for you to keep it the same way even at the end of the season. An easy technique for enriching the soil and keeping it protected after the season is over is to plant some green manure crops in the soil during fall and then mix it in with the soil once spring arrives. Examples of crops that are excellent for this purpose are cover crops such as alfalfa, ryegrass, and clover. All of these crops are green manure crops that can very well improve soil condition and structure and provide the beneficial microbes in the soil with nutrients, resulting in a richer and healthier soil for the next gardening season.

PART III

Fruit Gardening in an Urban Home

F ruits are popular among everyone, children, and adults alike. As a matter of fact, most adults develop a fondness for fruits right from their young age. We all love to have a supply of our favorite fruits in the refrigerator, but the problem is that most fruits are costly to get from the grocery store. Fruits are even more expensive when they are organically grown and out-of-season. As a result, a lot of people are not easting as many fresh fruits as they should be consuming. The amount of fruit on your daily table likely doesn't count towards what nutritional experts will call a healthy diet. Also, most people don't even know that there are wider varieties of fruits that are grown in the world. A lot of people haven't seen rare fruit combinations such as white currants, red currants, hybrid berries, and many others. Most of these "rare" fruits are fruits that are quite expensive in the grocery stores. Still, they can be easily grown by any homesteader with sufficient fruit gardening knowledge.

Still, ask someone – anyone – about the possibility of starting a fruit garden in their home. You will likely get, "I don't quite have space," or "Oh no, the fruits will end up being eaten by the birds." Some might even tell you that they'd rather have ornamental plants in their home because fruits are much trickier to grow. The fact is that all of these reasons are excuses, and they are wrong. People don't usually know just how much they are missing out on until they start fruit gardening themselves. As a newbie, it is perfectly realistic and possible for you to successfully build a garden of your favorite fruits, including apple, plum, cherry, peaches, apricot, raspberries, strawberries, blackberries, blueberries, gooseberries, figs, pears, and currant in your small backyard space. As you can see, the list of fruits you can have in your backyard garden is quite impressive. Save the exception of fruits like avocado, banana, and citrus; any homesteader would have happy to source for these fruits directly from their own backyard.

Growing fruits in your backyard is a way of encouraging your family (if any) to appreciate the appeal of nature. Fruit gardening isn't just about the pleasure you get from starting your own micro-urban garden; it is also about opening up an arena of learning for your young children.

An aspect of fruit gardening that many homesteaders appreciate is the fact that growing fruits present them opportunities to help bees – together, the fruit crops produce tons of flowers that crave pollination so badly. And these flowers look so pretty that they actually beautify the look of your homestead. Growing fruits in a backyard garden don't have to seem like hard chores; you will be fine as long as you know what to do and how to do it. If you want, you can plant your fruits in the same garden as your vegetables, or you can have a different fruit garden. This largely depends on the kind of vegetables and fruits you want to grow. Although some veggies and fruits go well together, some just can't survive in the same garden.

The best and most recommended way to grow your fruits is organic. There is little point to having homegrown fruits if they aren't produced organically, with little to no inputs and without the addition of pesticides, herbicides, or fungicide to the gardening mix. Organically-grown fruits are usually more delicious, healthier, and safer to consume than the store-bought and commercially-prepared fruits. When you have your own fruit garden, you can just choose to pick one fruit and pop it straight into your mouth, even without washing, because you know the process that went into growing that fruit and you are confident that there are no nasty chemicals that could affect your health in it.

Generally, everyone agrees that fruits are much more difficult to grow than vegetables. However, this is usually the case when you grow your fruit trees, bushes, and canes on a large scale. Fruit gardening also tends to be difficult when you grow the fruits too close with one another. Nothing attracts pests and diseases to fruit trees and bushes more than proximity. The pests may end up becoming more trouble than a newbie knows how to handle effectively. Growing your garden fruits on a small scale is the best way to go since this makes them less likely to attract plant diseases and pests. Know that this isn't a way of deterring you from fruit gardening. Instead, it is a way to let you know what lies ahead of your fruit

215

gardening journey. In any case, if you stick to the guidelines offered in this book, you will be able to start your own fruit garden without too many hassles.

Well, the first big step in fruit gardening is choosing the fruits you want to grow and the amount you want. You have your favorite fruits, and each member of your family has theirs, too. You may enjoy yourself some apples, while another person in your family loves apricots. The best thing to do when selecting the right fruits to plant is to ensure that there are great companion vegetable plants that you can grow together with these fruits. Soon, we will come to how you can choose the best fruit types for your new garden.

When you want to choose the right spot in your garden for growing fruits, the most important thing to consider – as you should already know now – is the soil. Fruit gardening is not much different from vegetable gardening; the key difference is the plant you want to grow. Naturally, you can follow the same steps for building your vegetable garden, but there are slight differences. Usually, the average garden soil is perfect for successful fruit gardening. But no matter your soil type, adding a bit of compost or any other organic matter always boosts the health of the soil. Fruits can be easily grown in many parts of the United States and the United Kingdom. However, some crops tolerate wetter soil conditions better than others. Regardless, the ideal gardening spot for your fruits should be in the full glare of the sun. But if you want your crops to have some shade, you can definitely arrange that.

The ideal growing soil for fruits should be around 18' deep and should be well-drained. Also, the soil should have a minimum of ten percent organic matter if you want to succeed; this is to ensure fertility. The best soil combination for growing fruits is either clay-loam, sandy-loam, or pure loam soil. Just make sure there is enough loam in your soil mix. Heavy, light, poor, or shallow soils can be easily improved and enhanced with organic manure and compost that can help increase the depth, structure, drainage, moisture retention, and fertility. The soil for your fruit garden should have a slightly acidic pH, and the ground must be ridden of perennial weeds before you even start planting.

From the above, you should be able to tell that choosing the right spot for your fruit plants requires you to leave out the wet areas. While waterlogged soil may be good for growing some types of vegetables, it rarely ever works for fruits. But, of course, fruits like cranberries can tolerate conditions like this. Similarly, you must also avoid areas of your garden that are too dry. There has to be a balance between moisture and dryness in the area where you want to grow your fruits. Unless you have an irrigation system installed, you should never plant your fruits in dry parts of the garden. If you want a low-maintenance garden, you don't want to invest all of your time in watering the fruit plants. A drip irrigation system installed around the roots of the plant can go a long way in minimizing the amount of time you spend on labor.

Most fruits need a sunny and sheltered spot for them to grow healthily. However, some fruit bushes like currants are appreciative of some form of shade. Regardless of what you do, protect them from frost pockets to prevent them from being damaged by the freezing cold in the early part of the year. You need to ensure that your fruit plants are in an environment protected from the biting cold to allow them to flower well. A protected environment makes pollination by insects easy, and it also allows for the fruits to set successfully. If you live in a windy area, it is best to grow a tall hedge to serve as a boundary or some sort of screen protecting your fruits. Or, you can take advantage of the fencing in your backyard homestead to make a major difference in the lives of your fruits. It's all about the choice you make, really.

Before you even get to planting, you have to create an actionable plan. Overcrowding is a real temptress when it comes to both fruit and vegetable gardening. Beginners often end up cramming as many fruit bushes and trees as they can into a miniature garden all because they want to reach a specific goal. Unfortunately, this always leads to the poor growth of all the plants involved. This is why you must have a plan in place. You can make your plan using a garden planner, as mentioned earlier. Use the planner to pick out fences and boundaries, and work out how many fruit trees and bushes can fit in the garden. As you add more bushes and trees in the planner, a grey circle pops up around them to show you how much space the roots will require. There are different methods that you can use to plant and grow your fruits, and they will be discussed in the following chapter.

In the meanwhile, below are some tips that will prove very useful in your fruit gardening adventure:

- Always start your gardening journey with easy-to-grow fruits such as strawberries. Strawberries require less room and maintenance, and they often produce very quickly.

- Make sure that your fruit garden is as small and simple as possible when you are getting started. You can expand the gardening area, the more experienced you become with fruit gardening and backyard homesteading in general.

- If you don't have a space that is large enough for vegetables and fruits. Try growing some of your chosen fruits in containers. There are different varieties of gardening containers that can be purchased from your local or online store for gardening. Otherwise, you can make your gardening

219

containers out of unwanted household objects and some garden objects such as wheelbarrows and old water butts.

- You can train up some of your trees to the fence, walls, or trellises in your backyard homestead. This is a very effective way of gaining space in the garden.

- Raised bed techniques are better for planting fruits such as strawberries, bush fruits, and canes. They are also good in gardens with poor under-area soil.

There are many other great that can really impact your fruit gardening success, and they will be revealed as you read this part of the book further.

Chapter Seven

The Best Types of Fruits
for Home Gardening

Naturally, there are several types of fruits that you can grow in your home garden, but you can't grow them all. Even if you have a small backyard, there are still tons of fruits that you can grow without restrictions. But before planting starts, it's best to put some thoughts into the fruits that grow the best in the climate condition of your state or local area specifically. Fruit trees and bushes can love for as many years as possible, and they require topnotch soil, proper sunlight, and air circulation. Fortunately, some fruit types just go in any garden. These are the best types of fruits to consider having in your garden.

Blueberries are some of the most popular fruits in a backyard homestead; you will likely find them at any backyard garden you visit and for good reasons. Blueberries are very easy to start growing, which is why many new homesteaders often try their hands at growing them as soon as they start fruit gardening. With their attractive shrubs, white flowers, summer fruit, and the red foliage, it is hard not to like blueberries or want them in your garden. Growing blueberries require a soil that is rich and acidic enough to feed them. You may need to do some form of advance work to make the soil acidic enough. The shrubs of a blueberry plant can easily stay alive and produce fruits for many years. To get a good harvest of blueberries at the season-ending, you need to combine two varieties to encourage good pollination. In the colder winter months, grow highbush blueberry varieties like Bluecrop. If you live in a place where the climate is mild, go for the southern highbush rabbiteye blueberries. You may grow your blueberries directly in the soil or in gardening containers. Just ensure that you cover the plants with netting so that they are protected from hungry birds when they start setting fruits. Blueberries usually require full to partial sunlight exposure for survival, and they require a soil mix that is incredibly rich, acidic, mildly or eagerly moist, and well-draining.

Strawberries that are freshly picked from the farm are usually worth the effort that you put into growing them. You have three options for your strawberry plants:

- June bearing which is great for planting in June as a large crop but has a shorter fruiting season;

- Everbearing, which produces three to four smaller harvests each season

- Day-neutral, which continuously produces small batches of strawberries throughout the whole growing system.

Generally, strawberries spread through runners. But if you want to have great production, your runners shouldn't be more than a few plants – prune the remaining plants. Also, be sure to pinch away the blossoms in your first planting season to prevent it from setting fruit. This is a way to help it channel its energy towards building a well-protected and healthy root season. After the first season, the production in the next season will increase to a significant extent. Finally, make sure you change or swap out your strawberry plant every 3 to 5 years Or, at the very least, make sure that you rejuvenate the plants in this timeframe. Strawberries need full sunlight exposure, and the soil for planting should be very rich, a little acidic, mildly moist or set, and be well-draining.

Apples are difficult to grow than most crops usually, but many homesteaders still want to have them in their garden. A basic idea of fruit gardening will tell you that apple trees are often affected by different pest and disease problems. New apple cultivars may be grown to be hardy. Still, you also need to protect them in your own ways, by covering, spraying, and other protection techniques that should be incorporated swiftly, specifically as soon as planting starts. Apple plants also tend to require a lot of pruning, which is another reason why homesteaders are wary of growing apples. To grow apple, you need two apple tree varieties to achieve pollination. Choose trees with several varieties that are grafted into a trunk, or go for a columnar tree, which can be homegrown in your containers. If you have limited space, you can plant the dwarf apple varieties. Pruning and thinning is essential for promoting good health and disease prevention in your plants. So, ensure they are a part of your routine when you are caring for your fruit garden. Apples grow best under full sun exposure, and you want the soil to be rich, mildly moisture as well as have poor well-drawn.

Raspberries and blackberries are favorites among homesteaders. For years, they have always been a part of most homegrown fruit gardens. But older varieties of raspberries and blackberries usually grow to be rambunctious plants, which means they quickly spread around the garden and become covered in thorns to the point that harvesting becomes a significantly difficult and painful task. Newer varieties are much better because they behave better, and they grow to be thornless. Moreover, the best idea is to plant a mix of the early, mid-season, and late-season raspberry and blueberry varieties as this helps to extend your harvest for many more weeks. These plants need you to prune them annually to keep them protected and productive, but this task is usually quite easy. The aim of pruning is to thin out the plant significantly to allow sunlight and air to reach every part of the plants. This promotes growth and helps with disease and pest control. In USDA growing zones, you will need 5 to 8 raspberry and blackberry for your backyard farm. The ideal sun exposure for these fruits' ranges from full sun to partial shade.

Grapes are another incredible fruit that should be a part of any home garden. Normally, grapevines are easy and straight forward to grow. Still, competition often comes from birds and other farm animals who want to have a taste of your grapes. Plus, you need to build support – sort of like a trellis – for grapes to grow on. Also, there are different instructions on the best way to prune grapes. Still, most gardeners can grow them successfully without using an aggressive pruning approach. Before you plant your grapevine, make sure you check with the local extension office in your local area to get valuable information on the best grape varieties to grow. And make sure you check whether the variety that is recommended to you is good for winemaking or eating. Most eatable grape varieties need to be located in a sunny part of the garden with rich soil, good drainage, and sufficient air circulation to prevent possible disease attacks. The right exposure for grapes generally is full sun.

Cherries are quite easy to grow, care for, and maintain. Unlike many fruits, they require little pruning sessions and very rarely suffer from pests or disease attack. To cross-pollinate, cherries must have two trees. Otherwise, you will need to plant one tree with two different grape varieties on it. Moreover, if you are focusing on just spur baking cherries, you can get away with planting just one tree. The best time to prune cherry trees is during the winter while it is inactive. Then, you should fertilize and add organic matter in early spring. Since cherry trees are not known to be drought-tolerant, ensure that you water them at least weekly in the hot months and let the rainfall take care of them during fall. Cherries need full sun exposure to grow and mature as you want.

Peaches tend to have very small trees that will fit right in any backyard farm, no matter the size. When peaches start ripening, their sweetness can be smelled from yards away. Additionally, one of the benefits of growing peaches yourself is that you get to enjoy the freshness straight from the source, instead of the stale and potentially-damaged options that are sold at the supermarket. To keep the branches of your peaches' tree productive and manageable, you will need to do some pruning from time to time. Thinning young peaches tree helps them produce small-sized crops of large peaches, instead of heavy crops of tiny ones. Peaches require full sun exposure every day to get all the nutrients they need.

Figs are another fruit that can be grown directly in the ground or in a gardening container. They are generally pest-free and are known to require very little pruning. There are different fig varieties that you can get to plant in your garden, and many of them are known to be quite hardy. In fact, some of the newest varieties are even hardier than the rest. If you want to grow your figs in a gardening container so that you can keep it indoors during winter, make sure the container is as small as possible. The less space between the roots, the smaller the treetop will remain. Plus, a small container is much easier to move around than a big one. And regardless of the size of the container, you will still get enough fig fruit to feast on with your family. Figs need full sun to partial shade sun exposure to grow in the right condition. Plus, the soil mix for growing figs should be rich with organic matter, moist, and loosely-draining.

Melons are the go-to fruit for the new homesteader that isn't quite ready for the commitment required by fruit trees, bushes, and shrubs. This is because they can be easily grown in containers or in the garden if it is what you wish. Melons require a lot of sun exposure and warmth as part of their growing condition. They also require ample spacing, as they usually grow on vines, which can easily grow up to 20 feet and more. You may be able to grow your melons on a trellis, but only if you purchase a variety with small-sized fruits. Watermelon varieties often become so heavy that they end up dropping off the plant. Start planting your melon right after the final frost, and make sure you keep them watered regularly to establish their growth. Once the fruits start to appear, you can ease off on watering. Melons, regardless of their varieties, require a full sun exposure to grow healthily. Additionally, the planting soil must be loamy, well-endowed with nutrients, and well-drained.

Companion Planting

Companion planting is a method of planting that you should really consider as you start to expand your gardening. This method of planting goes beyond the idea that some plants can benefit others when they are planted in proximity to each other. Companion planting is the planting of two or more crops of the same species together to control pests better and increase the yield. Although companion planting is mostly used for vegetables by many farmers, it also applies to fruits, and it is something you should consider using with your backyard garden. Many fruits grow amazingly well when they are planted with other crops. Below is a list of some fruits that should be in your garden and suitable companion plants:

- Blackberries – strawberries, dewberries

- Cucumber – peas, beans

- Figs – mustards, dandelions

- Grapes – blackberries, peas

- Apricot – chives, leeks

- Melon – pigweed, chamomile

- Peach – Basil

- Raspberries – tansy

- Sweet corn – pumpkin, squash

- Strawberries – onions, lettuce

- Tomatoes – pepper, cabbage

Of course, many other fruits can be grown alongside their companion plants. Companion planting is a really effective method that new gardeners can use to minimize the risks of pest invasion on their new farms.

Now that you know the exact types of fruits that you want in your backyard garden, let's get to the part where you actually plan and plant the fruits.

Chapter Eight

Planning and Planting Your Fruit Trees

Fruits usually grow on trees, bushes, shrubs, and canes. Depending on what you want to grow, you can buy your trees, canes, or shrubs from any outlet in your local area, including nurseries and garden centers. However, it is best to purchase from nurseries as they mostly sell organically-raised plants. Before you buy any of the plants, ensure that you look into the varieties. You should only get fruit varieties and rootstocks that are suited for your garden's growing conditions and needs. A great place to start is to check out the seed and plant catalogs. In most catalogs, there is usually a small session dedicated to fruits, and most of them offer a selection of the most popular fruit varieties. Specialist nurseries offer a broader berth of fruit varieties; so, you may want to contact them directly for advice and information on the catalogs. A lot of nurseries have catalogs that are displayed on their websites, and you can easily browse through these at leisure. If you have kids, you may want to invite them to the browsing session and also when you are planning and choosing plant varieties. You can even take them to the nursery when you go to purchase the fruit trees and plants. This can give a sense of anticipation and appreciation for what you are all about to achieve.

When buying your fruit trees, bushes, or canes, make sure to verify that they are disease and pest-free. This is highly important. Most people who sell fruit trees and plants have to register their stock with the local plant health agency, which helps to ensure that the plants are certified healthy. Usually, the healthy plant agencies can check and verify health and wellness for all types of berries, and they may be able to certify some of the tree fruits too. Unless the fruit plants come from a registered originator with verification and all, you probably shouldn't buy your fruit plants from them.

If you buy your fruit trees, shrubs, or canes from a specialist nursery, you will get them as bare-root plants – without soil, but packaged in damp peat and a protective sacking. They may also come as damp roots in a protective polythene bag. They are usually packaged and sent in this way in the dormant season, which is between November and March. Strawberries are packaged slightly differently. They should be planted as soon as they arrive, but if your soil is too dry, wet, or frozen, put the plants in place without frost and let the roots stay moist to prevent them from drying up. Before planting, you should soak the roots in a bucket of water for at least one hour. Fruits grown in containers can be bought directly from a garden center and be planted at any time of the year, as long as the soil condition is superb. Make sure you water the plants for one or more hours before you even plant them. Again, this will prevent the roots from getting dry.

How to Plant Fruit Trees Directly in the Ground

Examples of fruits that grow on trees include apples, pears, peaches, and cherries. They are all traditional orchard trees that require you to know about pollination, pest control, pruning, fertilizing, and other care tasks. To minimize the risk of disease, search for new varieties of disease-resistant apples. When planting fruit trees, it is good to plant dwarf trees that are close enough to the ground for you to pick their fruits. This is to make harvesting easier for you. By doing this, you won't need to lug around ladders or support yourself on them while working on your garden. A good advantage of dwarf trees is that they are quicker to bearing fruits that full-size trees. If your backyard is small, a dwarf tree is the best for you since it won't take up as much space as a full-size tree. If you are growing peaches, you can get a super-dwarf tree that will be easy to grow in a container or pot. Super-dwarf trees are miniature trees that are not more than 5 feet tall. Although many fruit trees can come as super-dwarfs, peaches have the most flavorful taste, and you don't need more than one tree to grow your own peaches. Plus, they are very easy to grow for beginners.

- Choose a suitable spot for planting your fruits.

- Mark out the positions where your fruit trees, bushes, and canes are to be planted. Then, use a tape measure to size out the planting distances and a market to draw the layout.

- The soil for planting should have been prepared for at least one month in advance. Dig a deep and large hole that is about one square meter in size. Continue digging deep until you reach a layer of topsoil with a light texture. Dig around the surface of the subsoil layer to break it up, mixing in a layer of garden compost. Make a slight mound at the bottom of the planting hole to help the bare-root trees position better and have something to support them.

- Get rid of any large rocks and weeds in the ground before you start planting.

- Bury in a strong, supporting stake (if needed) and drive it in so firmly that it remains sturdy in the ground without moving. Bare-root trees should have their stake driven vertically, leaving it on the side of the strongest wind (south-west). Container-grown trees should have their stake driven in from an angle where it won't be in the way of the plants' roots when planting starts. Drive the stake into the soil on the side to the strong wind to make it lean towards the wind. You will also need to install supporting structures for the espaliers and cordons before planting.

- Place the tree into the hole you dug and keep turning it until you are satisfied with the positioning. The tree stem should be positioned about 8cm from the stake. Fruit trees that are to be grown as cordons should be positioned at a specific angle.

- Make sure you don't plant the tree in too deeply – the graft between the upper part and the rootstock should be positioned above the upper level of the soil. If you have an old soil marker on the trunk, let this serve as the depth. Don't plant the trees too shallowly. If you have roots poking out onto the surface level, it means you need to dig your hole to go deeper.

- Carefully check the roots of your bare-root trees and gently trim the edges to get rid of roots that are damaged or over 30cm long. This makes them much easier to plant. Also, you should spread out the roots evenly within the hole, and then back-fill it with a mix of topsoil, compost, and any organic fertilizer. Be sure to shake the tree gently to spread the soil around the plant roots and firm it with your hands. After back-filling, carefully firm the tree using your foot, but make sure you care very lightly.

- For container-grown trees, you should carefully cut the root ball away from the container before you put it in the planting hole. Adjust the depth with the required planting mix, more or less. Gently, push part of the roots from the root ball and back-fill the planting hole with the soil mix, exactly as described above.

- Once the planting hole is filled to satisfaction, gently firm it for the final time. Then, make a shallow depth around the tree base for a basin that can retain water and soak it into the root ball. Water the tree some more and securely tie the tree to the stake to keep it in position. If the comes with a fairly long trunk, use a second tree tie to fix the base of the stem.

- If rabbits are likely to be a problem in your farm, protect the trees with galvanized netting or install rabbit guards around the ground where the trees are planted.

How to Plant Fruit Trees in Containers

If you have limited space, you can grow dwarf trees or super-dwarf trees in small pots or other containers. As mentioned above, fruits like strawberries are great for container-gardening. To plant your fruit trees in containers, follow the steps below:

- Choose a plastic, wooden, or terracotta container that is about 40 cm in size and put a layer of gravel in the base to serve as drainage, before you fill the containers with your planting soil mix. Preferably, you should use a multi-purpose compost that is heavily mixed with garden compost and topsoil.

- Plant the trees, bushes, or canines in the depth of the marked soil area, excluding blackcurrants which shouldn't go more than 5cm deep in. If necessary, you should put in a stake to serve as support. Use anything as thick as a bamboo cane.

- Water the container regularly – daily in the hotter months – and feed the plants every two weeks while the growing season is one. Tomato feed is great for fruit plants. Don't let the container become waterlogged – put some bricks under the container to manage the drainage if needed.

- If the container-garden produces a lot of food, you will need to get rid of some to disallow the branches from breaking. It is best to prune the immature baby fruits early in June before they become too big. Pinch extra fruits and let there be just one or two fruits on each cluster. If you are growing some gooseberries, know that they benefit from being thinned in late May, if you would rather have some really big fruits.

- Prune the trees regularly according to the instructions that will be provided soon.

- Protect your terracotta pots from the freezing winter by either moving them into a greenhouse or shed or by wrapping them up in bubble-wrap. Otherwise, they will end up breaking or shattering.

The Guidelines to Pruning Your Fruit Trees

Pruning may seem like a daunting task, but that is normally far from the truth for those who know the best way to go about it. Your fruit trees and bushes will produce better fruits so long as you prune them regularly. Pruning is unarguably demanding, but it is just as satisfying and rewarding for gardeners who take it seriously. The main aim of pruning is to encourage the production of more fruit branches and prevent unwanted fruit growth. Trained pruning goes a long way in determining the shape and size of your fruit plants. Pruning need not be complicated with the basic facts that are highlighted below:

- Use sharp secateurs to make a swift and clean cut on the plants without leaving any rough edges.

- When you prune, make slanting cuts around the surface of the upward-pointing bud. If you are pruning to return the plant to a vegetative bud, a new shoot will appear on the side that the bud is pointing.

- Ensure that you prune at the right time of the growing season, based on the types of fruits you are growing. Most gardeners carry out their pruning duties during the dormant time of the season, which is between November and March. However, you may need to prune some of your fruits during summertime to eliminate unwanted plant growth.

- It is quite easy to tell new wood apart from the fruiting wood on fruit trees. Just check to see if the fruit buds along the fruiting wood are fatter and rounder than the vegetative buds. This is much easier to notice in March when the buds start to swell. So, if you aren't sure, just wait until March before you prune.

- Prune away all damaged, diseased, and dead wood that is on the fruit trees to avoid the spread of pests or diseases.

As you can see, pruning isn't as complicated as you think. And it is certainly more important than you think. So, ensure that you get pruning done when needed the most. This is essential to your success as a backyard fruit gardener.

Tips for Taking Care of Your Fruit Trees and Bushes

- Water the young fruits once every two weeks. Make sure that the water you apply goes at least 3 feet into the soil. This is usually the depth where most fruit plant roots are extended.

- Fertilize the fruit trees once each season.

- Prune the fruit trees with your shears and secateurs.

- Use preventive orchard care to keep the trees free of pests and diseases.

- Treat any disease or pests that occur using the guidelines provided in the last part of the book.

PART IV

Raising Livestock

~~~~~~~~~~~~~

Raising livestock in the backyard of your urban home isn't something you should jump right into, especially if you are new to backyard homesteading. But it is definitely something to expand to once you start getting the hang of the whole homesteading thing. The point here is that you should focus on your fruit and vegetable gardens for a while before you bring farm animals into the picture. If you live in the city, you are not allowed to keep large livestock such as cows, sheep, pigs, and other animals that are too large to be contained in your backyard. These animals are outlawed in most local areas. If you want to raise large-sized livestock in an urban setting, goats are the closest chance you will get to achieving that. Unless you have a land that is more than an acre, at least, you shouldn't raise any farm animal that is bigger than a goat. Before you purchase your farm animals, it helps to check with the local municipal planning department to find out the specific rules guiding each species of animal you have in mind. These rules vary significantly from city to city. While some cities are very lenient with the raising of farm animals, others are more stringent in their approach.

Fortunately, the need and concentrated effort to reintroduce more natural and eco-friendly perspectives to people in the urbanized parts of the United States have allowed the creation of new regulations allowing small backyard farms and animals to live within the city limit. This is an excellent thing for the natural environment. Still, it is also a way for you to provide a direct means of survival for yourself.

As briefly implied, the first important thing any new farmer must do before they start raising livestock is to research the zoning laws and ordinances in their local area. You need to find out the number of animals you are allowed to keep. You also need to know the kinds of livestock that are permitted in your region. More importantly, you have to research the survival requirements of the animals – particularly their living facilities, feeding, and overall care. Can you take care of the animals as required and deemed necessary? In most states, you may be asked to submit certain documentation to the city before you get permission even to purchase and start keeping a livestock farm. Some laws monitor how you can sell the eggs or milk that you get from your livestock.

There are several farm animals that most regions generally allow people to raise in the backyard of their homes. Below are some of the most popular ones to consider adding to your homestead.

Chickens

Chickens are unsurprisingly popular with backyard farmers. Keeping chickens in a backyard farm gives you unrestricted access to fresh egg supply and meat, which means you will never have to go to the grocery store. Apart from the fact that they are good sources of eggs and meat, chickens are also affable companions – which explains why many farmers cannot resist the endearing cuckoos of a flock. Don't be surprised if you start keeping some as pets after getting your first flock; chickens are that easy to get close with.

There are two different types of chickens, and both types perform different functions on the farm. Firstly, you have layer chickens that are kept exclusively for their ability to produce an amazing amount of fresh eggs every summer, and sometimes in the winter as well. Then, you have broiler chickens that are kept and raised for their heavy and delicious meat. Modern chicken breeds are being bred to excel at either of these purposes: egg-laying or meat supply. However, chickens generally have edible meat, and they produce eggs, whether they are layers or broilers. The main discerning thing between layers and broilers is the rate at which they produce eggs. If you want to keep chicken primarily for their eggs, you can get a flock of egg-laying breeds, and vice versa. But you can also get dual-purpose breeds that are sure to satisfy you with both eggs and meat.

You can easily get your flock of chickens from any online hatchery where they have hundreds of breeds that can be mailed to you for overnight delivery. The breeds of chicken you choose should be based on the egg-productivity rate, growth rate, looks, and ability to resist diseases. Most people don't like to slaughter birds that they put so much hard work into raising themselves. If you fall into this category of people, you can find a mobile abattoir, if any in your neighborhood, and let them do the work for you swiftly, for a fee. Ensure that the abattoir you go to is permitted to do the butchering by your local regulations, though.

The best chicken breeds with super egg-producing abilities usually lay at least 5 or more eggs every week of summer. Still, this number may fall down to two, one, or nothing in the colder months. Granted, there are ways to ensure that your chickens keep producing eggs even in the colder climate. One of these ways is to install artificial lighting in their coop to complement the insufficient warmth from the daylight. Evaluate the rate at which your family consumes eggs to decide on the best numbers of chickens to keep in your backyard farm.

Chickens need a well-ventilated and moisture-free coop with one nesting box for each bird to survive in your backyard farm. They also need you to provide them with roosting bars, which is where they will lay their heads each night. Also, you have to build a run with enclosed wire mesh where they can forage, scratch, and dust-bath as much as they want. You need to let the hens out of their coop every day at dawn and then let them in once it is dusk.

Note that you have to consider your chickens' neighbors' convenience when you start raising them. For obvious reasons, you shouldn't keep roosters in your flock unless you want your hens to reproduce. Most cities don't allow roosters, though, which means you won't be able to keep a rooster even if you wanted to. Also, you will need to clip your chickens' wings so that they can't fly over the garden fence to destroy your neighbors' beautiful garden. The only exception to clipping their wings is if you plan to keep them enclosed in a specific space all the time.

Turkeys

Turkeys may not lay eggs as often as chickens do, but they produce some incredibly delicious eggs that will leave you wanting more at all times. Like your backyard chickens, you need to ensure that you have the proper housing and spacing to keep your backyard turkeys in. You can raise turkeys for their eggs and meat, just like hens. If you want chickens that will serve as a reliable source of meat whenever you need it, you should go for heritage turkeys since they are widely-acclaimed for their wonderful taste and amazing plumage.

Turkeys are primarily meat birds because of their inability to lay eggs at the same rate as chickens. As a new backyard farmer, you should avoid the modern commercial turkey breeds. These come in two, which are, broad-breasted white and broad-breasted bronze turkeys.

They are known to be highly susceptible to diseases and infections, plus they find basic bird functions, such as walking, difficult. Heritage turkeys are much better in meat quality than the other breeds. You can get heritage chickens from any online hatchery. If you don't have a local abattoir in your residential area, you should reach out to the local poultry processions to find someone that will be willing to accept the turkeys in small batches.

If you want to raise some chickens, you must have at least two in your backyard farm. You should never raise a single bird alone because they require the company of one another in a flock. Raising two turkeys means you get to keep one for Thanksgiving and the other for Christmas. If you want to have enough turkey meat that can stay in your freezer and last you throughout the year, know that each turkey will produce up to 15 pounds of meat once it reaches the age of maturity.

Similar to chickens, you also need to build a survival coop for your turkeys, with the roosting bar and run. But you may strike out the nest boxes since they won't be laying eggs too often. You will need to clip the wings of your turkey birds as well to prevent them from flying over the fence.

Ducks

Want more birds in your backyard farm? Consider adding some ducks to the collection. Depending on where you live, free-range ducks start from a dollar upwards. Naturally, ducks don't produce quite the same number of eggs as the average hen, but they are surely close in yield. In fact, some homesteaders argue that some breeds of ducks can out-perform most chickens in egg production. But that isn't the only reason to love ducks. Many farmers also agree that ducks are better garden companions than chickens and turkeys. Regardless of their purpose, most ducks usually produce an impressive amount of meat and duck fat. Some duck breeds, such as the Rouens, Khaki Campbells, Indian Runners, and Blue Swedish, are good small-scale birds that are easy to raise in a backyard farm. If you want a dual-purpose duck, consider going for the White Pekin, which is acknowledged as an excellent source of both eggs and meat. To make your ducks happy, purchase a kiddie pool and some great feed.

If you don't want to purchase a kiddie pool, you can build a small water pond for your ducks to play in. But understand that kiddie pools are a better alternative because of the low price. Provide your ducks with a freshwater fountain and a secure home where they can be protected from predators. The feeding requirements will change for the birds as they mature, but be careful not to overfeed them to avoid an overload of protein. This can be really harmful to their health.

Rabbits

For whatever reason, many homesteading farmers aren't raising rabbits as they should. This is kind of a mystery because rabbit meat is actually very delicious. Although some people just use them for breeding or fiber, the most profitable part of rabbits is their meat. However, I feel like a lot of people don't like to raise rabbits as farm animals because they can't bear the thought of raising such cute bunnies and then hauling them off to the abattoir for slaughter.

Having said that, young rabbits can be butchered as early as 12 weeks, which is even faster than you can for chickens.

Raising rabbits in the backyard is great, but you can also raise them indoors. They are great animals to keep if you have a fruit or vegetable garden as their manure is so rich and organic that you don't have to compost it beforehand. They are great companions if you want an animal that can also serve as your pet. Plus, they are ideal for you if you live in an area where you can't keep other livestock in your backyard. Rabbits are generally edible, but you should avoid those that are bred for the show alone as they don't have enough meat on their bones. Unlike chicks and poults, rabbits aren't available for mail delivery from suppliers. So, you have to research if there is any local breeder in your area. If any, purchase a few rabbits from them to start your backyard rabbitry.

To raise rabbits, you have to have a keen sense of organization, but caring for them is a relatively easy task. Plus, the investment that goes into keeping rabbits is less costly than the amount you spend on larger animals such as chickens and goats. They need smaller fencing, less feed, smaller pens, and less expensive breeding stocks. Moreover, they can be raised in a very small part of your backyard. With the right plant, you can raise hundreds of rabbits every year in your backyard garden. However, you may not need as many like that to feed the family. '

For breeds that are raised for meat purposes, consider breeds such as New Zealand, Crème d'Argent, California, etc. For pet breeds, go for the Lops, Miniatures, or Dutch rabbits. But of course, if you don't have an interest in their meat or company, go for breeds such as the Angoras; their hairs can be mixed with fibers for blending purposes. To start raising breeds, you need to start with one female and one male; then, these will start producing more offspring than your family can consume in a year. To prevent overbreeding, don't allow your rabbits to breed unless you are running out of rabbit meat in your freezer.

Taking care of rabbits involves letting them out of their hutch into the run every morning and then closing the hutch up right before dark. You also have to provide them with sufficient food and water. Weekly chores include a thorough cleaning of the hutch every week. Compared to other livestock, rabbits are very clean and quiet animals. So, you don't have to worry about them disturbing your neighbors.

Goats

Goats may not be acceptable backyard animals in your local area, so make sure you check for validity before you bring them home. Goats are raised for their meat and milk, but it depends on the breed you get. Meat-breed goats don't usually produce enough milk for a family; so, if your goal is to get milk from your goats, be sure to go for dairy goat breeds. These kinds of goats usually don't produce as much beef as the meat breeds, but they are quite edible too. Eventually, you will have extra goats on your hand that can either be eaten or sold. Female goats rarely produce milk until they have reproduced and given birth. For an urban micro-farm, I highly recommend dwarf goats, which are the generally-acceptable type of goats to keep anyway.

Goats are very social animals, so don't get anything beyond two. You likely won't have space for more than four to five goats anyway. Two mature dairy-breed goats are all you need to get the required amount of milk for an average family. Plus, there's usually plenty of milk left to make cheese and yogurt. Goats are okay with being kept in any three-sides building, shaped like a shed and lined with straw. You will also need to build a large fenced area where they can enjoy their own company. Ideally, you may need to place a few big boulders for them to clamber around in the fenced area.

You should feed your goats and milk them at least twice a day as part of the care routine. Their nutritional need isn't that difficult – just make sure that they have access to good pasture or bales of hay, at the very least. Also, make the thorough cleaning of their pens a weekly task and trim their hooves at least once every six weeks. Goats are extremely curious animals, though, which means you need to have a fence that is sturdy enough to prevent them from barging in your neighbor's flower or vegetable garden. Mature male goats are usually not suitable for city life; so, you may need to take your female goats to any goat farm in the nearest countryside to you so they can breed. Keep in mind that goats also produce excellent manure that will prove very useful for your backyard fruits and vegetables.

If your city allows you, other livestock that you can raise in your backyard include cows, pigs, horses, and other large animals. If you want, you can even add some bees to your farm animal collection.

Chapter Nine

Housing and Spacing Guidelines for Your Backyard Animals

Housing is the most basic requirement for farm animals. You shouldn't leave your animals out in the open at all times. There should be a shelter where they can lay their heads at the end of each day. Milk-producing animals such as goats or cattle are more vulnerable to cold temperatures than other animals. Without proper housing, their teats can become frozen, which halts milk production completely or, at least, slows production down. A proper animal shelter is key to keeping your livestock happy, relaxed, and productive. Farm animals require some form of shelter to keep them safe from the many elements that are out in the open. Unfortunately, many homesteaders make the mistake of thinking that the open backyard is enough shelter for their livestock – this is wrong. Some even believe that the only time to provide shelter for livestock is during the winter months. But the fact is that animals can tolerate cold even better than humans can, thanks to their natural coats. Summer heat is way more unbearable for most livestock and farm animals. If you don't provide shade for your animals during hotter temperatures, they likely won't forgive you, and this will be obvious in the way that they relate with you.

The kind of building you provide for your animals will depend largely on the climate of the area where you live. Shelters come in different forms. We have barns, pole sheds. Tree belts, thickets, fences, and many other types of shelters that farm animals can be kept in. Regardless of the type of housing you want to build, the shelter should be built according to the sizes of the animals. If you have plenty of animals of different species, you have to provide adequate spacing to avoid overcrowding and disgruntlement among the animals.

Whether it is the structures you build for them as their homes or the trees around your backyard, there should always be a source of shade for your farm animals. If you do build them some structures to live in, you have to ensure that there is enough ventilation in the buildings. Many farm animals, such as rabbits, don't have to sweat, which makes them very vulnerable to heatstroke. A three-sided shed with an open door or front is often the choice building for most animals that live on pasture, e.g., goats. When designing the three-sided shed, make sure the front is in the south direction, away from the strong winds. Build the shelter on a well-drained, elevated, and a leveled ground where you can easily feed and provide them with water.

When constructing a shelter for your livestock, you should keep some factors in mind. One of these is the air quality of the building. Sheds and shelters for farm animals should have natural ventilation, which is why you are advised to keep the front open. If you want to keep the shelter enclosed, make sure there are enough fans and air inlets to help air circulate better in the building. Tight buildings without proper ventilation often leads to a concentration of animal odors, and respiration gases, which can cause damage to your livestock's lungs and lead to pneumonia. It may also trigger the instant increase of the ammonia levels, which can result in the suffocation and death of your animals.

You should also keep the drafts in mind when building your livestock shelter. Animals may be able to withstand colder temps than humans, but this doesn't mean you shouldn't shield them from the cold drafts. Building strong panels in front of the open shelter greatly reduce the build-up of drafts. Think of drafts at animal height, instead of human height. When you allow your livestock to roam freely around in pen instead of hitching them, they will naturally seek out the most comfortable spots to settle down in the backyard shelter.

The sleeping and bedding area for the animals should be kept as dry as possible. Animals are far more comfortable during winter months when they have dry beddings that are also clean. Thick and dry bedding offers warm from the cold ground. It reduces the amount of body energy that the animal has to expend to keep itself warm. Shelter away from the rain and snow also allows the animals' coats to stay dry, which further increases their insulating value.

All farm animals need a regular water supply, probably more than they need their feed. Water is essential to the survival of your livestock. In cold temperatures, make sure there is a regular supply of fresh water. If not, make available a freeze-proof watering device to keep their thirst satiated. The animals are more likely to drink when you keep the water at 50-degrees Fahrenheit. Otherwise, they may suffer from dehydration without you being aware of it. Adequate, fresh food is just as vital to your animals' survival as the water supply. Animals are better at managing colder temperatures when they have enough food to produce energy and then convert the energy into body warmth. Animals also require the energy which comes from their foods for proper growth and maintenance. Therefore, you need to provide them more quality feeds during the cold climates. For the herbivores, make sure you supply them with purchased feed and free-choice hays.

Spacing

Depending on the size of your backyard, you need space to provide your animals with exercise yards and pastures. If you don't want to practice pasture at all, you will need to buy enough feeds to keep them going, set aside an exercise yard, and create a good plan for effective manure management. If you do make use of pasture, the number of animals that the pasture can support will depend greatly on the soil mix fertility and other environmental considerations. The environmental considerations vary across the country. Using the rotational grazing method prevents the animals from overfeeding on the available pasture, prevents pest and parasite overload, as well as supports more animals than other systems, such as a set stock system.

Chapter Ten

Taking Care of Your Livestock

Caring for your animals is not just about feeding them and providing them with regular water supply. There are other aspects of animal care and maintenance that you must take seriously if you want to be successful in your animal-rearing venture. As explained, adequate water supply is vital to the survival of your animals. A few guidelines have already been provided on how you can make water available to your livestock. However, there is one very important guideline that comes in handy during the colder months. In the winter months, it helps to heat up the water that you give to your animals. Giving livestock cold water while they are battling with the chilling temperature lowers their body temperature, causing them to burn more body fats to keep themselves warm. Inadvertently, this also means that you will need to feed them more so they can keep replenishing their energy. The appropriate amount of clean and fresh water is key to your livestock's health and survival. Water helps to reduce or eliminate the risk of colic or impaction.

Of course, you already know that you have to feed your animals with the most nutritious foods that will boost their health and growth.

Farm animals require a balance of much-needed nutrients to maintain their health. The foods you provide them should have a nice blend of the right minerals, protein, and vitamins that will provide them with all the energy they need. Nutrients like these are best gotten from organically-blended feeds more than forages, but that isn't to imply that foraging doesn't play a part in their health. During winter, you should up the food intake of your animals. The lower the temperature, the more foods (energy) that the animals will need to conserve and regulate body health. Make sure you monitor the individual food intake of each animal to avoid overfeeding or underfeeding any of them. This is especially important with chicken flocks where they eat in order of the hierarchy established in the pecking order. Feed your animals with small amounts of food every couple of hours to reduce the production of water, regardless of the number of animals that you have on the farm.

Large quantities of mud/manure in the animals' housing quarters can make them highly uncomfortable. The situation gets even worse when you are in winter. A shelter with soiled beddings is a recipe for disease and sickness for the animals. Do not leave manure and mud to build up to the point where it can affect the health of your animals. Use tile, wood chip, and fresh sand across the floor of the animal shelter where and when necessary. This will help make sure that you don't have to deal with more waste than you can handle.

Whether you are raising livestock as a source of food for your family, as a hobby, or as a source of income, these caring guidelines will help you keep things in perspective.

Water isn't the only thing that will help keep your animals comfortable during the summer months. Certain types of shelters are very good for dealing with extremely hot weather conditions. The ideal types of shelters during extremely sunny temperatures are those that protect your livestock from the sun and provide a cooling effect to help them retain water in their body.

Constructed shelters are effective for keeping farm animals protected during the summer when the weather is unkindly hot. Shelters like this are building using materials that range from shade cloths to timber. The roofs are ideally made with galvanized steel or aluminum in many cases. Constructed shelters are good for reflecting the radiation from the sun. Trees that have large canopies also serve as effective shelter for your animals during the hot weather. Naturally, trees have this cooling effect that makes their sheds extremely soothing for animals and humans alike. This is due to the absorption of heat by the leaves on the tree. During the sunny temps, it is important to direct wind flow towards the animals to keep them feeling cool. So, keep this in mind when you are pondering on the location and type of housing to build for your animals. Making sure that your farm animals stay cool during the hotter climates is a key part of livestock care.

Unless it is totally unavoidable, you should never handle your animals when they are on heat. If you absolutely have to, make sure you do it during the time of the day when the temperatures are at their lowest. This is usually early or late in the day. According to research, handling livestock such as goats or cattle during hot climate conditions can hike their body temps from 0.5-degrees to 3.5-

degrees, which is quite a lot. The animals may be unable to handle the temps at that level. Increased heat stress and temperatures can result in the loss of productivity in the livestock and may even affect their normal functions. On the other hand, moving animals during the colder weather conditions can reduce the impact temperature has on production performance.

If you need to transport your animals during extreme climates, you have to consider their welfare. The transport of livestock during heat or extremely cold weather can adversely affect the health of the animals; therefore, it should be generally avoided. However, if you absolutely need to transport them and you can't put it off, you need to map a journey plan that minimizes the effect of the weather on their health. The most basic step to take is to predetermine the route you will be taking for your journey. Also, mark out the possible stops along the way and see if there are places where you can stop to get shade and water. If you need to stop along the journey, park your vehicle under the shade and make sure you are at the correct angles for the wind flow to come towards the animals. Keep the stops to a minimum so that you can get to your location as soon as possible. Heat builds up even more when a moving vehicle remains stationary for a while.

Some animals are at high risk of heat stress than others, and you should keep this in mind as you care for your backyard livestock. Younger animals, darker-colored animals, and sick animals are at more risk of experiencing heat stress. The ability to tolerate heat stress usually varies from specie to specie. For example, goats are less prone to heat stress than animals such as pigs, llamas, etc. Dairy goats are, however, more susceptible to heat than meat-breed goats. The examples go on like that among the many farm animals that can be raised in a backyard homestead. Watch the animals that are more prone to heat stress closely to see if there are any signs of heat stress during the hotter climates.

How do you identify heat stress?

There are many signs to watch out for in your animals, but some of the most general ones that will show in most farm animals include:

- Increased breathing rates

- Panting

- Increased intake of water

- Appetite loss

- Increased salivation

- Lethargy

- Loss of consciousness, in extreme cases of heat stress

If any of your livestock is exhibiting symptoms of heat stress, some of the things you can do to help them cool down include:

- Move them under the shade immediately you notice. Choose a spot with a breeze. If the animal is too stressed to move from its spot itself, find a way to provide shade in that spot.

- Offer them a lot of clean, cool water and encourage them to drink in small amounts. Spray their coat with cool water, particularly on the legs and feet, or let them sit or stand in a pool of water.

- Increase air circulation in that area. Do this with ventilation, fans, and anything that can make air move more freely.

- Reach out to your veterinarian if the animal (s) does not show any sign of improvement.

Although heat stress isn't completely avoidable, its effects can be decreased to a reasonable extent by following the caring guidelines and tips above.

Chapter Eleven

Protecting Your Farm Animals
from Predators

A ny good farmer that cares about their animals knows that predators are a problem that must be taken care of. Naturally, you should be worried about predators and the damage they can cause to your livestock if you aren't careful. You should also be wondering about the steps that you can take to keep your livestock protected from predators. There are different ways that farm animals can be protected from predators. Still, some methods are more straightforward and easier than others. In this chapter, we examine some of the easiest and most effective ways of protecting farm animals from predators. There are a variety of options for you, so feel free to choose one or more of the methods. The good thing about the methods is that you can combine two or more together to give your animals the tightest security possible.

The first effective method is to use guardian animals, such as dogs, to protect your livestock. Guardian dogs have been around for years, and they were developed via selective breeding of specific dog breeds in Europe and Asia, specifically to protect livestock from wild animals, such as bears and wolves. So, yes, there are dogs that can be dedicated completely to the protection of your backyard animals. Of many guardian animals, dogs are considered the most effective, and for obvious reasons. They can protect livestock from a comprehensive range of predators across all sizes. If raised and trained properly, guardian dogs can help decrease or completely erase predation on your farm. They can even help your livestock add more weight (and meat) since the absence of predators helps them graze and forage more comfortably. However, not all dogs are guardian dogs. Therefore, make sure you ask specifically for a guardian dog breed when looking up dogs for your animals' protection. Getting any dog that hasn't been specially trained to be a guardian dog can cause a lot of problems for you.

Fencing is another efficient method of keeping predators away from your livestock and your gardens. Coyotes can be an extremely troublesome concern for you if you decide to add some sheep to your farm. Adult coyotes can squeeze through 4-by-6 inches woven wire, and they can also jump over fences that are below 66" in height. This means that it is almost impossible to build a completely predator-proof fence. Some homesteaders have successfully learned to add electrified wires to their fences to keep predators from entering their backyard. Others simply increased the amount of trip and top wires in their fences, and still went ahead to electrify the fence. From extensive research among successful backyard homesteaders, it can be said that electrified netting is the best way to keep predators out of the backyard using your fence. Although fences with electrified netting certainly don't last as long as wire fencing, they are inexpensive in comparison and can reduce predator invasion from 47 percent to about 6 percent. That is a whole lot, as you can see. Another way to use fencing to keep predators away from the farm is to use a fladry. Fladry involves rigging a fence with electrified wires, as well as flags. However, fladry is more effective for smaller areas, such as a calving pasture. It is best combined with a strategic grazing position. So, it isn't exactly the best fencing option for a whole backyard farm. But it can help with small areas in the garden.

Making regular changes in animal husbandry and management is another way you can protect your livestock from predators. Essentially, this means that you adopt farming practices that reduce your animals' exposure to predators. For instance, you should always move any dead animal far away from the garden to reduce the possibility of a scavenger being drawn to the herd. You can also adjust the times of the year when you lamb, kid, or calve. The weather in April to May isn't only considered pleasant, it is also the time when many wildlife animals are giving birth to their young ones. The more opportunities for them to focus on something other than your livestock, the better for you and the animals.

Making management changes also involve changing the schedule of your animals, i.e., the time they go out of their shelters, the time they eat, and the time they return to the shelter. You may also need to change the type of oversight they receive while they are living their lives. Livestock management isn't quite an easy task. If it were, we would all have a couple of cattle, chicken, goats, turkey, and other livestock that we are managing in our backyards. But if you want to be a successful homesteader, it is something that you have to care about and devote your time to. Suppose you need to revisit your schedule to manage the predation concerns. As an example, you can consider moving your livestock to another part of the backyard if there is enough space. You can change their feeding and foraging schedule, and you can update the technology installed in your farm so that you can manage your livestock better. This includes installing alarm systems and setting traps with trip wires to catch predators and warn them from returning to your property.

PART V

Effective Pest Prevention and Control in Your Fruit and Vegetable Garden

---

Pest problems are frankly annoying. A seemingly bountiful season can be ended so abruptly all because of nasty pest infestation. Dealing with pests is a natural and irremovable aspect of gardening, regardless of whatever it is you are planting. Even the best gardeners experience crop failure occasionally. It is impossible to find a gardener or farmer that will tell you that they have never had a problem with pest – if you find any, makes sure they tell you their secret so we can all apply it to our gardens. However, managing pests in your garden may be just as easy as it is for the pests to take over. And it is unbelievably fun and enlightening, but only if you prevent them before they take over your garden. Prevention is better than cure. Once the outbreak happens, pests become much more difficult to manage them effectively. Even if you are finally able to contain the outbreak, the damage to your crops would have been done. And, you would have wasted a whole lot of time, money, effort, and other resources that go into planting and growing fruit and vegetable gardening. The very thought of losing your beloved crops can be very frightening.

One thing many homesteaders don't know is that some organic pesticides are just as harmful to growing crops as the chemical-based sprays. It is better to avoid using pesticides for your gardens, organic, or otherwise. The point of pesticides is to kill insects that are benefiting from the growth of your plants. Killing insects is the purpose of applying pesticides. However, they often end up altering the pH levels of the soil. They also leave toxic residue on your crops. That's not all – pesticides can also destroy the beneficial microbes that are found in healthy soil. Or, they can wreak a combination of these three negative effects. One of the common pesticides used by many gardeners is the soap-and-water spray that has been shown to have the potentials to kill microbes and disrupt soil pH levels, depending on the dilution of the solution. If you don't want to damage your garden's ecosystem, it is much better to prevent pests than fight them.

A vital way of naturally preventing pests in your fruit and vegetable garden is patience and endurance. Instead of going straight to pesticides when you suffer pest infestation, it sometimes helps you to wait for a little and continue following all of the natural pest preventive techniques that we will soon discuss. While you continue to do your part, your soil microbes will keep learning and familiarizing themselves with the new environment. Eventually, a balance will be achieved. You are probably wondering what the connection between beneficial soil microbes and pests is. Well, soil microbes are the ones that feed your plants, helping them remain healthy and well-protected against pests. When you spray insecticide, it affects the pH level and disrupts the establishment of your soil's beneficial microbes. This means that you won't get the balance you crave, which further means that the microbes won't be able to feed your plants as healthily as they should. From there, you will develop a never-ending reliance on pesticides because you will keep using them if your microbes can't effectively keep the pests away. Patience is usually the key in situations like that.

To prevent garden pests from destroying your fruit and vegetable garden, here are some of the tips that can help. With these tips, you won't have any need for organic or chemical-based insecticide sprays.

- The first big step is to improve the quality and health of your soil. Healthy soils give plants a healthy immune system. With a stronger and healthier immune system, plants become much better at fighting diseases and pests. Healthy soil is key to feeding your plants with beneficial microbes.

- Some plant varieties are more resistant to pests and diseases than others. Choose fruits and vegetable varieties that are naturally pest-resistant. The catalogs usually contain information on the varieties that are best at resisting diseases and pests.

- Make sure you choose the perfect spot with the right amount of sun exposure and everything else that your plants need. If your plants need full sun, then make sure you plant them in the full glare of the sun. Likewise, you can plant your fruits and veggies according to their water requirement. If one plant needs a lot of water to keep its health, then you should plant it the spot that retains moisture better than the others.

- Encourage beneficial insects to come to your garden. Beneficial insects are any insects that consume the pests around your plan. They naturally come to gardens themselves in the search for nectar, pollen, and shelter. Encourage them to keep coming or stick around by growing flowers that keep attracting them and meet their needs. Once beneficial insects know that they have a habitat in your garden, they will start laying eggs to expand their army and start getting rid of pests from your garden.

- If you plant strong-scented herbs with your vegetables, they can be very effective for getting rid of pests. It is a super-easy and effective way of preventing garden pests while also helping yourself with the herbs. Some of the best strong-scented herbs that you can plant alongside your vegetables are garlic, coriander, calendula, etc. These herbs should be planted at the edge of the garden so that they can prevent the pests from getting near the crops.

- Practice crop rotation. Using crop rotation to grow your crops confuses pests and improves soil fertility in the garden. Crop rotation is challenging to use in a small backyard effectively, but it isn't unachievable. If some pests overtake a plant in a particular spot, don't plant it in the same spot the following two seasons. Or, you can use a cover crop to enable that crop to rest for one season. Crop rotation is difficult but efficient in preventing pests.

- Interplanting is another great preventive method that focuses on alternating certain crops, flowers, and herbs to confuse the pests. Naturally, pests love to attack mono-crops, which is why commercial farms have to use pesticides heavily on the plants. Instead of planting your garden with mono-crops, alternate each row of vegetables with a row of insect-attracting flowers and pest-repelling herbs. Confusing pests is a sneaky way to hide your crops away from the strong nose of the pests.

- When you still have young plants, use floating row covers to keep pests away from them. Light-weight row covers allow light and water to penetrate the plants while keeping the pests from them. If you experience a recurring pest problem with a particular vegetable, you should consider using low tunnel hoops to protect that crop. But ensure that your life your row covers away each morning so that pollinators can find their pollen.

- Build pathways in the garden. Permanent pathways are effective for encouraging beneficial insects to come to the garden since the temporary ones which you have to till each year usually end up destroying the insects and their habitat. The type of material you use for your pathway will depend largely on the situation you are in. Gravel, white clover, and wood chips are, however, some of the materials that you can use. Building permanent pathways mean that you will also have permanent garden beds that will continue to increase in fertility over time. Of course, fertility is one of the major factors that determine your garden's ability to repel pests.

- If you have a few pests in your garden, don't get rid of them. Sometimes, having some pests in the garden actually works to your advantage. It may seem counterproductive, but without some pests to keep them in your garden, the beneficial insects that are meant to feast on the bugs won't stick around. Beneficial pests are only attracted to gardens where they can find bugs, their favorite foods.

- When the few pests in your garden turn into an outbreak, immediately get rid of the infested plant to keep the damage under control.

- Be proactive instead of being reactive in your approach to pest management. Being proactive means taking steps to prevent the infestation of pests instead of waiting for the infestation to happen first. A possible pest outbreak is a potential way for you to learn about ways you can strengthen your ecosystem. For example, you can observe your soil and try to determine if it is missing a vital mineral that is leaving the plants sick enough to attract bugs. If so, is there any organic material that can effectively substitute that mineral.

Track the pests that you encounter in your garden, the time they appeared, how you tried to manage them, and the results of your preventive and management techniques. This can open your mind to the aspects you want to focus on the journey to preventing pests from infesting your garden.

Chickens and other livestock animals are also very effective for controlling pests in the garden. If you are tired of pests destroying your plants, just set your animals on them. The birds in your farm aren't just there so that you can eat them or their eggs; they can also make a difference in your efforts towards keeping pests at bay. Chickens, turkey, ducks, etc. are charming bug and slug control. Using your farm animals to get rid of pests is a cost-effective option; it is also chemical-free, which means that your plants will stay safe regardless.

Poultry birds have a very keen sight that allows them to see bugs everywhere in the garden, field, or lawn. They can even spot rodents and snake and also get rid of them. Guinea fowls, in particular, are great for this purpose. Not only can they spot bugs, slugs, rodents, and snakes from afar, they also have an alarm system to alert you when you have intruders on your garden. Guinea fowls consume anything from the most vicious disease-bearing bugs to the gentler ones. To make it better, they can also serve as a border patrol for your farm. So, if you don't want a guardian dog to monitor your animals, you can get a guinea fowl to perform that same task. But only if you are sure that you (and your neighbors) can handle their loudness. Guinea fowls are known to be very noisy, with the females being even noisier than the males. Male guinea fowls usually won't call out to you unless they have an absolutely important reason to do that. Once the male calls out, though, the rest of the flock starts chiming in, until they create an intimidating alarm that should bring you immediately to the farm.

Preventing pests from taking over your garden and jeopardizing years of hard work is a vital part of your gardening journey. Remember that prevention is better than cure so, instead of waiting until you are hit with a pest infestation, start taking proactive steps to minimize the risks and possibilities. This will help you far better than any combative method.

Chapter Twelve

Harvest and Preservation

So, you have successfully planted your fruits and vegetables, tended to them, and now, here comes that time when you get to reap the fruit of your labor. How do you go about harvesting?

Before you harvest your vegetables, make sure that they are really ready for harvest. In an earlier chapter, I gave instructions on how you can determine when it is harvest time for your vegetables. If you have ascertained that harvest time is right, then start picking the fresh vegetables. A big mistake that many veterans and new gardeners make is that they wait for their vegetables to become really old and mature before they start picking. You should never wait until your produce is too mature, old, or big. If you do, the produce might turn bitter, tough, and even damaged. Check your vegetable garden every day to see which of your veggies are ready to be picked, preserved, and stored away. Frequent harvesting is a way to up your vegetables' productive abilities. So, pick your vegetables while they are still young and tender. Remember than homegrown backyard vegetables rarely get as big as the ones you buy from the markets, considering that you won't be adding any chemicals to speed up the growth.

Always harvest when you see that the plants are dry. Harvesting wet produce can trigger a spread of disease, especially on some vegetables. The best time to harvest is early in the day when your produce is hydrated. Once you harvest, you should consume the vegetables within two to three days so as not to reduce the freshness. However, if you would rather store the vegetables for later use, then you should use the guidelines below to preserve and store your veggies.

Most fruits and vegetables will store for months if you harvest and preserve them the right way. The key to successfully storing your crops is to keep them in the right conditions and check on them regularly to rid them of any diseased items. The specimens need to remain unblemished throughout preservation. For example, one bad apple can ruin the rest of the batch if you don't regularly check on them to remove the rotten and damaged ones. Storing your crop in a well-ventilated and dry area can prevent rotting from even taking place. It helps to purchase storage boxes where you can keep the crops until you are ready. However, wooden crates and shallow cardboard boxes can work just as well.

To preserve your harvest, there are many methods you can use – depending on the types of crops you are preserving. Freezing, canning, drying, and pickling are all effective ways of preserving your fruits and vegetables, as well as the herbs if you decide to grow any. The method you use from these four techniques will be determined by what you want to do with the produce when it is time to consume it. For instance, if you harvest some blueberry bushes and you would like to store them away to make blueberry muffins later, freezing the blueberry bushes is the right thing to do. But if you would rather turn the blueberries into a jam to be used throughout winter, canning the blueberries into one or more jams is likely the best approach you can take.

Freezing is an efficient way of preserving a variety of fruits and vegetables from your garden. While it is a relatively simple method of preservation, it entails so much more than tossing your vegetables into an air-tight bag and then throwing the bag in the refrigerator or freezer. Before you freeze your veggies, it is best to blanche them. Blanching your vegetables means cooking them briefly in boiling water. This should be done to vegetables such as tomatoes, peas, sweet corn, and beans. Freezing your vegetables as soon as you harvest them from the farm is the best thing to do. Blanching your veggies before you freeze them helps preserve the original color of the veggies, reduce the loss of beneficial vitamins, and cleanses the vegetables. Blanching should be skipped when you are freezing any fruits. Note that some vegetables just don't like it when they are frozen. These include cabbage, cucumbers, and celery. So, don't bother freezing them to avoid creating waterlogged messes. Frozen vegetables can last up to a year in the freezer if you keep it at zero degrees Fahrenheit.

Drying is known to change the taste of the produce. This is due to the removal of water, which leads to the concentration of the flavor. Drying also changes the texture of the vegetables. Peppers are examples of vegetables that are best stored by drying. You just need to find a dark, cool area where you can hang the peppers up. It helps to let the vegetables take a little bit of heat when you are drying them out to store away. Drying your vegetables involve four steps:

- Prepare the veggies by thoroughly cleaning them and removing any debris or dirt.

- Blanch the vegetables.

- Dry the vegetables

- Store in airtight containers and put away in the closet or pantry

Monitor the veggies as they dry up, as they may become too dry and possibly burn if you let them dry for too long. You can tell if a vegetable has dried up once it becomes flaky and crumbly. Peppers and tomatoes become crispy when they are ready to be stored.

Pickling is mostly used to store cucumbers, but you can also use the method for carrots, cabbage, peas, and beans. There are two ways of pickling: refrigerator pickling and fermented pickling. Below are the steps to refrigerator-pickling your veggies.

- Sterilize the pickle jars by putting them in large saucepans, covering them with water to the brim, and put on the stove. Let the water boil to sterilize the jars.

- Chop up the vegetables you want to pickle into sticks, slices, or any other shapes. Make sure they are all of the same sizes. Fill the jar with the vegetables.

- Pour a cup of vinegar into a small saucepan and add a tablespoon of salt. Wait till it dissolves, then add a cup of water and set down from the stove.

- Add seasoning to the jar where you have your vegetables. It can be anything from garlic clove to peppercorns. You may also add a tablespoon of pickle seasoning.

- Pour the vinegar solution into the vegetable jar and let the veggies be completely covered. Wait till it cools off before you cover and store in the refrigerator.

Let the pickles stay in the refrigerator for a couple of hours before you try one. You should have yourself a true pickle, but the taste will keep developing over the next couple of weeks. Refrigerated pickles can last for a few months.

Canning is the method of preserving your veggies by making them inaccessible to bacteria. Any vegetable can be canned, from your tomatoes to blueberries. However, canning is the most complicated of all preservation methods that have been discussed so far. There are so many things to consider when you are canning your vegetables. Only can vegetables that have been harvested straight from the garden. Glass canning jars are the best for storing the vegetables; make sure they aren't cracked or chipped. Before you can any of your produce, check to make sure that the cans are fully sealed.

Conclusion

Backyard homesteading is a very sustainable and self-sufficient way of improving your quality of life by producing the foods you eat yourself. From your favorite vegetables to fruits and meat, there is nothing that you cannot effectively grow and raise effectively in a backyard homestead. This book, as promised, provides the best information, guidelines, techniques, and methods of keeping a fruit and vegetable garden, while also raising livestock, all in an urban homestead. Without further ado, the only thing left to do is to get right to backyard homesteading and start putting everything you have learned to use.

BACKYARD HOMESTEADING

Growing Flowers and Beekeeping in an Urban House

MONA GREENY

Introduction

A ges ago, self-sufficiency was the only way to live; modern society, industrial machinery, and many of what we take for granted today were non-existent back then. That said, if you track homesteading through the years, you will always go back to agriculture. Even though it may sound odd for people in this day and age to look for sustainable, self-sufficient channels instead of opting for shopping from the store next door, it can come in handy in times of need and during crises. It's important to delve into the historical roots of homesteading and how societies viewed self-sufficiency on both micro and macro levels.

Many early political philosophers, like John Locke, have thought of homesteading as a way that can settle the argument of land ownership, doing labor on the land, in his view, would benefit the entire community. The mechanical and industrial progression of many countries has made them realize that a lot of vacant lands can be benefited from. Aside from common law, statutory laws formulated by the U.S. laid the groundwork for many major land acquisitions that benefited the people.

The Homesteading Act of 1862 is considered one of the most important and effective legislative acts that have changed the course of the United States for years to come. Signed under the harsh conditions of the Civil War, Abraham Lincoln knew that passing the law meant that individuals would be able to make use of over 270 million acres of public land, which amounted to around a tenth of the land of the whole U.S..

The requirement of being a homesteader, according to the 1862 act, was quite simple; the individuals in question would have to be 21 years old and were forbidden from taking up arms against the government. This allowed settlers from around the world to claim land, in addition to single women, former slaves, and other people whose rights were not in the best shape back then. To own the land, an individual had to build their home, make improvements, and grow a sustainable farm for more than 5 years. This process would happen only after the homesteader filed an official application, and their property was assessed. After 5 years of improving the land, the homesteader would finally receive their land patent signed by the president of the country.

The Homestead Act remained in effect for around 114 years until the majority of the prime land available had been homesteaded. The homestead certification or land patent, however, continued to be a proud sign of determination and aptitude. This old revolutionary concept still seeps into modern culture as more people are considering a self-sustained lifestyle instead of completely relying on fewer choices and less diverse alternatives.

In the '70s, movements like the Back to the Land are a stellar example of how the urban and suburban communities and culture were not focused on autonomy and self-sufficiency. Instead of being a means to survive, homesteading became a lifestyle that thousands of people practiced. As such, getting in touch with one's ancestral roots has become prevalent in Europe as well, whether it was for survival, distributism, or battling pollution.

While this outlook may seem political for those who only want to sustain themselves to lower their expenses in urban neighborhoods, understanding the true nature of these modern movements is just as important. Modern life has can be draining and can have limited options, making many people reconsider their position and take a proactive initiative. Things like government corruption, social stigmatization, and excessive consumerism have led to major landscape-changing events, including the Vietnam War, scandals like Watergate, and various types of pollution.

It's not really counterintuitive for people to start considering cutting off some of the roots that make their life increasingly difficult and under the control of business-oriented entities. The first groups of people who started reorganizing their life to synchronize with homesteading were those who were already familiar with farming, wildlife, and rural life. What started out as a simple movement in the late '60s transformed into a full-on philosophical and economical approach that whole countries are now trying to adopt. The sustainability problems that many countries face pose serious dangers to both the population and the environment, making it one of the top priorities of both developing countries and superpowers.

Those who don't have a lot of experience in farming, raising animals, or living in rural areas may be too overwhelmed to take such a decision into consideration, which is only normal. This is why a lot of people start resorting to things like backyard homesteading and gardening to provide the bare minimum that they need for themselves. The good news is that progress never stops; once you start sustaining yourself in one department, you'll want to move onto the next.

Homesteading is one of the best lifestyles you can adopt if you're looking to hone and improve your work ethic. It's not uncommon to see children at the age of 7 and 8 taking some responsibility around the house in rural homesteads, milking cows, training dogs, and helping with the cooking. This can be an important element that you may want to factor in if you're planning to raise a family with a strong work ethic and a sense of dependence that allows them to choose for themselves what they want to do with their lives, which is definitely better than standard and generic methods.

If you're someone who enjoys cooking their own food, you will definitely enjoy planting and growing your own vegetables. While you may not get used to not having the convenience of buying everything from the grocery store, adopting this lifestyle can unlock a lot of simple joys that you didn't know existed. It's important to understand that homesteading in this age is, fortunately, not a means to survive. You have the space that allows you to make mistakes and learn from them at your own pace. A lot of farmers in rural areas may feel the pressure to maintain certain crops to avoid seasonal changes that can stop them from having access to food year-round. You'll get to progressively choose your crops based on your needs and gradually make progress from there to be able to maintain sustainable products that allow you to be dependent on them.

Your perspective is bound to change once you actually start working for the things you've always taken for granted dramatically. This doesn't mean that you'll be suffering and wasting all your time planting crops and beekeeping, but rather appreciate them more as activities, and not just as means of sustaining yourself. You'll be able to enjoy the literal fruits of your work, which somehow makes everything you taste and see all the more worthwhile. You'll notice that everything you get out of homesteading have a longer lifespan and are utilized better, and you'll also be able to cut down on waste

Homesteading has a serious financial aspect that can often be overlooked if people obsess over the initial costs or labor. It may take you some time to notice it, but once you start realizing that you're becoming less dependent on market prices, you will discover that you're saving a lot of money. There will be no need to worry about centralized food supplies or markets that suddenly have their prices hiked. This kind of independence will save you a lot of expenses down the line, especially if you start thinking about straying away from the main power grid and establishing your own energy supply or source.

Unlike conventional homes, losing your job wouldn't necessarily be a devastating blow to your whole life. Sure, becoming unemployed would still pose a lot of problems, but homesteading ensures that it doesn't affect your food security and survivability. When you're growing your own food, a big portion of it isn't actually consumed, but preserved. It's not uncommon to find many homesteaders with a food supply that can last them months because they have preserved it in freezers, cupboards, and the pantry. Depending on the type of homesteading you're planning to do, food security is going to be the least of your worries more often than not. You don't have to think of extreme survival situations, but it's still reassuring to know that your backyard will be able to sustain you, even if you temporarily lose your source of income.

You'll be embarking on a journey that can show you the true roots of civilization, where you'll be getting in touch with the innate survivalist nature that we all have deep inside of us. All you need to start your homesteading journey is a modest backyard, which will surprise you as its true potential slowly unfolds when you put in the effort. This guide will mainly focus on the gardening aspects of homesteading, specifically when it comes to growing flowers, designing a garden, tilling, and taking care of it. The second section of the book contains detailed guides that should help a complete beginner start their own little beekeeping endeavor, coupled with all the details and information you may need on your beekeeping journey.

Chapter 1

Tool and Equipment

Getting into the world of homesteading isn't as complicated as it might sound, but it is not that simple either unless you are well-prepared. This isn't the kind of endeavor that you could take on haphazardly without any consideration or planning. Being self-sufficient is certainly worth the effort, but you have to put in the time and dedication so you could reap the rewards of your efforts. To get into homesteading, you need to have the right tools and equipment.

You need to start by setting a budget for each item so you're made aware of what might be off-limits and what is affordable. To get into homesteading, you will have to have some capital to purchase all the things that will help you on this incredible journey. In this chapter, we will explore all the tools and equipment that you will need for homesteading, so grab a pen and paper!

For the Garden/Backyard

Garden Fork

Most people who get into homesteading do it because they want to grow their own food, and that means you will be doing a lot of gardening and farming. To start the list, you need a heavy and reliable garden fork. This tool is particularly important if you haven't been paying attention to your space in a while, which probably means that you haven't had the time to improve the garden soil using friable sand and rich compost. The kind of fork that you should get needs to have 3 or 4 prongs (tines).

The gardening fork needs to be sturdy with a 44-inch handle shaft, and it should be made of excellent materials — pay attention to the type of steel used and make sure it is of high quality. Be careful not to confuse this fork with the long-handled one used for hay. A heavy-duty garden fork doesn't come cheap, but it is certainly worth the investment owing to its versatility and the number of tasks you can use it for.

Cart or Wagon

This is one of the most important tools for any homesteader. You should always prioritize getting a wheeled cart or wagon before any other tool. You will always be moving fertilizer, bedding for animals, compost for the garden, cleaning stalls, and a ton of other things. For those tasks, you will need a sturdy cart or wagon. It will take a load off your back, quite literally, and help you transport items around the backyard with ease and efficiency, not to mention swiftly as well.

There are many things to consider when it comes to a wheeling cart. First is the number of wheels, which will vary depending on the size as well as your needs. It is also imperative that you consider the materials that the cart is made of. The sturdier it is, the longer it will last and the more reliable it will be during your everyday tasks. Metal carts are ideal for heavier loads, and you will find that handling them is much easier when you're moving large or bulky items.

Tractor

A tractor isn't a must, but it definitely comes in handy if you are serious about farming and homesteading. The size and abilities of the tractor will depend on your budget, because those don't exactly come cheap. However, they are an excellent investment, nonetheless. A tractor will help you plow the soil, and will also come in handy in tilling, harrowing, disking, planting, and a host of other tasks. One of its main advantages is just how versatile it is because you won't only be using it in farming. A tractor can be used to pull other machinery and heavy items as well as pushing them, which proves quite useful for homesteaders.

Getting a powered hauler isn't an obligation for homesteaders, but if you invest in one, you certainly won't regret it. You don't even need a two-ton tractor; different sized machines could fulfill your needs without being too heavy or costly. You have cheaper and smaller options like gators or power wagons, which give you the advantage of a hydraulic machine to help you on the heavier tasks in the farm or backyard while being cheaper and easier to move around than giant tractors.

Lawnmower

Some people think that lawnmowers are a fancy accessory that just serves the aesthetic value of a backyard, but they serve a much more important function than that. It will help you maintain the length of your grass and weeds so that they don't spiral out of control and risk compromising your efforts. Some farmers use goats for that purpose in pastures and fields, but a lawnmower is a much more effective tool, especially considering that you are backyard homesteading. This makes it ideal for such spaces, while farms with bigger acreage utilize tractors for mowing the grass and weeds. Research the available options and find a lawnmower that caters to your needs and your backyard.

Cutting Spade

You will also need a sturdy cutting spade for your soil. It is one of the older farming tools, and its function has changed over the years, but it is still as important as ever for homesteaders. A reliable cutting spade will help you cut off sod or any clay clods that might be moved accidentally while using the garden fork. Those two tools work well together and can be used in a variety of tasks like prying rocks out of the soil and the likes.

Compost Bin

A compost bin is used around backyards or farms to make and house compost until you can use it in the garden. So, what's so special about it? The compost bin can accelerate the decomposition of organic substances by providing aeration and moisture retention, which means you get to use the compost when you need it without any delays.

Tiller

A tiller is one of the most important tools for any farmer, not just homesteaders. This machine is designed to break up hard soil that is compact and rough into loose dirt that you can use to farm or plant. There are front-tine and rear-tine tillers, and you would be better off with the latter. It should be made of cast iron, steel, and bronze for maximum efficiency. The rear-tine tiller will help homesteaders who want to experiment with organic gardening and incorporate natural soil enhancers like sand, manure, and compost in their backyards or farms.

For Woodworking and Repairs

Ax

A homesteader will do a lot of woodworking and carpentry, and you need an ax for that. It will help you cut, shape, and split barks, not to mention that it also comes in handy when cutting trees and harvesting the timber that you need for the woodshed.

Drill

A drill is an indispensable item for any homesteader especially one that likes to work with their hands. For woodworking, get yourself a cordless drill or driver that you can move around freely while in the woodshed. It will help you build anything you want out of wood, no matter your skill level. It makes it so much easier to handle wood and build structures out of it, big or small, which is why it is an essential tool. The drill or driver will also help you, should you need a tool to handle any other repairs around the house or the farm.

Hammer

You will also need a hammer because it is still the best tool to drive nails or stakes, and this will make it quite beneficial around a homesteader's backyard or house. It is also pretty useful for woodworking, especially for thinner or weaker wood structures you're working on where a drill might not be appropriate. A hammer is as good for driving nails as it is good for plucking them out, which might prove useful in repairs around your house as well.

Utility Saw

While a good old-fashioned hand saw is useful, and you should get one, a circular utility saw is something else and is much more versatile and useful. It is the perfect tool for any carpentry around the house, and it can be used in a host of other tasks. One great thing about the circular saw is the fact that the blades are easily replaceable, so you can replace any damaged ones or just put in new blades of different measurements for a certain task that requires this change.

Sawmill

A utility saw is excellent and can do a lot of the wood jobs you'll need around the farm, but if you plan on frequently using wood around your homestead, then a sawmill might be a better investment and the more suitable tool to use. It is much more efficient than a utility saw or even a chainsaw, and it can perform the job in a cleaner and smoother way for a better finish. This isn't the kind of item you have inside your house, though, so consider a sawmill only if you have a separate woodshed where you get most of your carpentry done.

Miscellaneous Tools

You should get a set of study pliers of different sizes because you will need them. Pliers will help you cut, bend, and handle wires, as well as a host of other functions. A set of screwdrivers is another necessity for handling any nails around the homestead, and they prove to be quite functional and useful. Speaking of nails, it is always a good idea to stock up on a lot of nails of varying sizes for repair jobs and any unexpected problems you come across in your homestead, and there will be many of those, so it's always best to be prepared.

A staple gun is another tool that proves pretty useful around a homestead, and it can save you a lot of time and effort in small tasks that won't be so small without this tool. A staple gun quickly puts things together, so it will help with roofing, fabric, wood, and anything that you want to put together efficiently and quickly.

You should also get different types of wrenches (pipe, combination, hex key, etc.), safety goggles, chisels, ladders, and any other tool that you might use in your homestead, even if you think you wouldn't use it that often. It's important to find anything you could need at hand so you wouldn't have to put off working on any projects.

For Raising Livestock

Raising livestock is essential if you really want to be self-sufficient, which is what homesteading is all about. If you want to be completely self-sufficient there are some things you will need to make raising livestock easier for you and the animals.

Chicken Feeders

Raising chickens is quite useful, whether you do it for their eggs or their meat, and they always prove to be a valuable addition to any homestead. A chicken feeder is important so that you can keep them well-fed. Most importantly, you will need a chicken waterer as well, because they must have access to clean and fresh drinking water to grow healthy.

Fences and Cages

Not all homesteaders raise chickens. Others prefer rabbits, for instance, and those need to be contained, or else they could easily flee. For other animals like cows or goats, you will need a solid fence to keep your backyard contained so that the animals wouldn't roam off, which can be dangerous if there are predators in your area. When you get a fence, make sure you also get the necessary tools to maintain it like fencing pliers.

Power

Many homesteaders are concerned about the environment. This is why they try using cleaner sources of energy, which is quite doable and even recommended for homesteaders. You might also find yourself backed into a corner if you are not near any reliable power source. One of the perks of backyard homesteaders is the fact that they are often in rural areas, which gives them more options for clean energy, although it can be done in more urban settings. In any case, for the average homesteader, you need to take a lot of factors into consideration and prepare accordingly.

Solar Panels

Solar energy is probably the best source of energy for homesteaders, and it is one that comes with a ton of benefits. You don't have to be near a power source, which, as mentioned, is a problem that many faces. The sun is always around, and you can utilize its energy to power your house and farm by installing solar panels. You must understand your homestead's power needs beforehand, so you know what size of solar panels you need and invest accordingly.

Solar panels are indeed an investment because they save you a lot of money in the long run. Moreover, they increase the value of the property because clean energy is the direction the world is headed in right now. Last but not least, solar energy is great for the environment and has a minimal carbon footprint, so it is ideal for homesteaders who want to live away from the pollution of the city and its energy sources.

Wind Turbine

Another renewable source of energy that a lot of homesteaders use is wind. This is particularly useful in areas where the wind blows hard, and you can harness that energy to power up your homestead. You just need to install a wind turbine, and you will find that it is quite practical and efficient in providing your house with the energy it needs.

Generator

Even if you rely on solar energy for power, which doesn't get affected by power blackouts, it is still a good idea to get a backup generator. You never know what might happen, and the last thing you need is running for a few days without power because of a storm or unexpected weather that might do something to your solar panels. With that said, it's advised that you invest in a heavy-duty generator that could fulfill your homestead's needs in terms of power should anything go south.

Stock of Fuel

Always stock up on extra fuel tanks, seeing as they are essential for a homestead. Whether it is for the lawnmower, tractor, generator, or any other appliance that runs on gas or diesel, tanks of fuel need always to be available, or else it might hinder your progress and negatively affect something you are working on.

For the Kitchen

For a homesteader, going to the grocery store whenever you need something isn't really an option, so you want to make sure that you have everything you might need. The most important tool for your kitchen might just be pressure canners because they will be used to store food that doesn't have high acidic content, and that includes anything from tomatoes to stews or rice. A meat grinder will also prove handy around the kitchen, considering that you will most likely process your meat, so make sure you get one if that is the goal.

You will need a set of knives of different sizes — make sure you get some big ones for cutting or skinning livestock. Mixers, meat slicers, solar ovens, meat saw, coffee grinders, and blenders are also some of the appliances that you might need in the kitchen of your homestead.

Before you begin working on building your very own homestead garden, it's essential to make sure that you have all the needed tools at hand. Some of these tools are more advanced than others, and may not be needed for a beginner, while others may already be in your kitchen or your handyman kit. You can browse prices online to get the best deals, especially if you plan on purchasing tools in bulk.

Chapter 2

Basic Setup

Now that you know the tools and equipment you will need to start a homestead; it is time to understand how you can have a basic setup and how you can get your own homestead up and running. It is important that you take things slowly in this step because a lot of details are entailed. You have to be organized and disciplined, so the setup phase can go smoothly and without any complications. Here's what you need to know.

Evaluate Your Needs and the Property

Before you start setting up your homestead, you have to take a moment to evaluate your needs fully and what you want to achieve through this experience. Homesteading isn't something to be taken lightly, and you can't just dive into this without proper consideration. A lot of hard work and money come with this step, so you have to be certain that this is what you want. Many people idealize the idea of homesteading without fully understanding just how much effort is required to make this work. This is why you have to be certain that this is what you, and more importantly, your family, want. Your partner and kids need to be on board, too, because they will be required to do work on the homestead as well. This is not suitable for single-member households.

If you've come to the conclusion that this is what you want, it is time to consider what degree of self-sufficiency you're going after. Some homesteaders just plant their own food, and they otherwise rely on external sources for anything else. So, you have to decide if this is something that would be convenient for your needs, or if you want to go full throttle and raise livestock while also getting your private source of power through solar panels or wind turbines.

While deciding on what degree of homesteading you're after, you have to take your property into consideration. Does its size make it possible for you to raise livestock? Can you install a renewable source of energy, or are you forced to resort to power stations? If this is a property you plan on leaving one day, this is something that you have to take into account. Will it be worth it spending all that money on wind turbines or solar panels, tractors, and all the other tools and settings if you're moving in a few years? These are questions that only you can answer.

Prioritize

Homesteading complications and difficulties aside, it is a very rewarding and exciting endeavor that will keep you occupied with physical labor. This is why it is important to take things slowly. You need to start by making a list of all the things you want to do on your homestead because you can be certain that you will be overwhelmed by ideas and things that you can do in every corner of your homestead. So, make a list of all projects and ideas you'd like to take on. This list will come to fruition after you evaluate your property in terms of size, soil, and capacity to accommodate animals, and based on that, you will set your goals. What kind of projects do you want to achieve in your homestead? Through this list, you will learn the answer to that question.

You can divide the list into different categories, depending on the type of projects you're interested in and which aspect of the homestead it pertains to. For instance, you can decide on which herbs, fruits, or veggies you want to plant. Then, you can decide what kind of animals you want to raise on your farm (if possible) and if you want them to reproduce or if you would just be raising them for meat and dairy or eggs. You could also set up a beehive for fresh honey. These are mere examples of the kind of things that you could do on your homestead, and the sky's the limit if you think about it.

Yet, you need to be very careful to avoid getting carried away by all those incredible ideas. You need to prioritize. If you try to do everything all at once, you won't be able to get anything done properly. So, you need to set your priorities straight and focus on the things you need urgently. This will most likely have to do with power or sustenance. You should always focus on getting those urgent matters out of the way before all else.

Start with a Garden

We will discuss the specifics of designing a garden and how you can do that later on in the book, but for now, you should know that the first thing you could start working on is this. This might sound daunting for a lot of people, but it just requires some planning, and you will be able to do it if you're willing to put in the effort. The first thing that you need to do is to learn what is suitable to grow in your soil and climate. You definitely don't want to waste time, energy, and money trying to farm some crops that have no chance in your soil type.

You can visit the local extension office to understand the planting calendar for your land — what crops can grow, and at which particular time of year. This step is particularly important for beginners in the world of farming and homesteading because there is a lot to farming that takes time for you to learn, and you need help for that.

Set Up an Irrigation System

For most people, the first thing they usually start with is a small garden or planting trees. For that to happen, you need to set up some sort of irrigation system around your homestead because you will probably be needing it in the long run. Irrigation systems are complicated, and setting one up entails an understanding of many intricate details, but let's dive into the general idea. Any irrigation system is divided into one of two categories: low or high-flow irrigation. As the names imply, for the low flow, water drips on your soil at a slow pace. As for fast flow systems, greater amounts of water are used, and the pressure is significantly higher, too. So, which should you go with?

The answer to that question depends on your preference and a couple of other factors. For starters, how big is the area that you want to irrigate? If it is just your backyard as a homestead, then you might be better off with a low-flow system. But if you're talking acres of land, a high-flow irrigation system is the reasonable choice because you can irrigate larger strips of land at a much faster rate. These are the two general types of irrigation systems in terms of speed and pressure, but there are other terms that you need to be familiar with:

Subsurface Irrigation is when the system is buried beneath the soil, you directly supply the plant roots with water. Examples of this system include drip irrigation and a soaker hose system, which are both low-flow options.

Localized Irrigation is a different system where the water is dispersed through pipes s, and this often happens above the surface. While this system is above the surface, in most cases, it is designed in a way that wouldn't flood the foliage, which helps protect it from mold and other complications.

You have several systems under each category, from soaker hoses to spray irrigation or drip irrigation, and it is up to you to decide which system would work best with your garden. In any case, you should read up as much as you can about those systems because irrigation is everything when it comes to growing crops properly. If you do anything wrong in this step, your crops can easily die out. It would obviously be better to consult with an expert on which system would work best for your garden, but you can figure that out yourself if you spend enough time understanding the subject.

Compost

Another gardening recommendation that you need to apply early on is composting. It is much better for the soil and will keep it significantly healthier, while it also produces less trash and waste. The great thing about composting is that it is not as difficult as you might think, and you can do just about everything on your own, even building your own compost bin! You just need a trash can, and you can add your coffee grounds, eggshells, leaves, and grass clippings. And just like that, you have your own composter with minimal expenses, and it can help your soil and increase its fertility in the long run, as well as reduce waste.

Turn Your Kitchen Around

The kitchen is one of the most important places for a properly functioning homestead, and you need to treat it as such. You will be making your own food and planting vegetables and fruit, and this needs special handling in the kitchen if you want those to survive. The first thing you need to learn in the kitchen is how to preserve food, because you're naturally going to have a lot of excess food, and it needs to endure time and conditions so you could use it later on.

The last thing you need is for your excess food and produce to go to waste because you failed to preserve them.

You're going to need to learn skills like dehydration, canning, pickling, freezing, smoking, cold storage, and a lot more if you want to keep your excess food intact. So, invest in a food dehydrator and a ton of canning supplies, they will prove quite useful and save you a lot of money in the long run.

Raise Livestock

Even if your homestead isn't that big, raising livestock is still a necessity and something you will have to do sooner or later. Unless you plan on growing only vegetables, you will need to at least raise some chickens. Their eggs will prove very valuable, whether they are for meals or for preparing other food. Moreover, chickens prove much more useful than just providing eggs or meat. Their waste can be composted and will help your soil. You might also consider quails, which don't need as much space as chickens and also provide eggs – – they're also silent, unlike chickens that do make their presence well known.

Chicken and quails aside, if you have space and suitable conditions, you would benefit from some meat. The great thing about the livestock you raise is the fact that you control the environment and what they eat and drink, so you will be eating the cleanest and best meat you could ever have. There are a lot of options for raising meat in your backyard, though cows and pigs won't work with every homestead. But you still have options like rabbits, for instance, which are a great source of white meat. In any case, you get to decide what kind of livestock you want to raise; just be certain that you can handle their needs and keep them healthy and well-fed.

Make Homesteader Friends

Contrary to popular belief, homesteaders aren't recluses who don't talk about their progress. You will find that many of them are friendly and quite eager to share their experience and what they have learned, and you should capitalize on that. Their knowledge can and will prove vital and will help you get on your feet and successfully manage your homestead. You will run into a ton of problems and complications that you most likely will struggle to deal with on your own, which is why mentorship from other homesteaders is very valuable. Any challenges you're facing, they've likely gone through, and they can show you how to deal with any problems that arise.

Technical experience aside, moral support is also pretty nice. Homesteading can present some frustrating challenges, and they might prove too difficult for some. Having a support system of friends who share your vision and believe in what you believe in can help you overcome such obstacles and push you forward. Moreover, having homesteader friends opens the door to trading with them and mutually helping one another. You might have a product that they want and vice versa, and you can deal with each other regularly, benefiting each other's homesteads.

Get Handy

If you're running a homestead, you need to become a handyman, whether you like it or not. Learning to build and repair is a skill like any other, and you too could learn it if you put your mind to it. Set up a small workshop where you can do any handwork around your set up, which will be quite often. Things break or malfunction all the time, and you need to be able to deal with those complications on your own. You wouldn't really be self-sufficient if you called a repairman or contractor anytime a bolt is loose around the house, or your pipes need a little tightening.

So, get the tools and start working. Try to do some building on your own at first so you could get the hang of things at first. Build a table or a cabinet, and experiment with wood or steel. The more you do this, the better you will get, and soon enough, you will know your way around a workshop like a pro. This will not only save you a lot of money in the long run but will also help you become truly self-sufficient and independent without being at the mercy of handymen.

Work Around the House

Being a homesteader isn't just working around the barn, garden, and workshop. There are quite a few things that you will need to be doing around the house, too. For starters, you should learn how to sew and fix clothes. The clothes you use in farming or repairing will wear out pretty quickly, and you need to be able to fix them because running out to get a new pair of pants or a shirt every time yours has a hole isn't really sustainable or self-sufficient.

You should also try doing things like making butter, cheese, or yogurt through raw milk. Try your luck with baking, because you will need bread and going to a bakery every other day isn't the homesteader's way. Learn to utilize whatever resources and tools you have around the house; everything is useful and can be used to accomplish something, even if you don't know how to do so just yet.

Set Up the Energy Source

We talked earlier about getting solar panels or wind turbines to power up your homestead, and this is the final step in completing your homesteading setup. Consult with an expert on this one so they could advise you on what would work best with your location, and then you can invest in solar panels or wind turbines, and make sure that the connections are made by someone who knows what they are doing since this will be your main power source.

Setting up your homesteading space may prove to be a hassle at the early stages, seeing as you'll want to focus on functionality, cost, and efficiency simultaneously. Most importantly, you'll want to opt for green options, so to speak, that will keep the environment unharmed and your crops healthy. If you're looking to save on costs, in the long run, you should always opt for natural energy sources, such as solar panels, to keep your homestead running at a low cost.

PART ONE

Growing Flowers

Chapter 3

Selecting the Ideal Types

To become a better gardener and homesteader, there are many things you need to familiarize yourself with before picking out the ideal flower types for your homestead. For starters, it's essential to understand the differences between perennial, biennial, and annual plants along with their pros and cons. This way, you will have a better chance of making calculated decisions to help in making your backyard homestead more abundant with varying flowers that serve different purposes, instead of wasting time and effort into growing non-beneficial plants. Many beginners focus all their effort and attention on growing food while neglecting flowers, seeing as they're always seen as merely ornamental. However, flowers do not only add a delightful touch of color and a sense of comfort to your place, but many of them also make the perfect low maintenance option when it comes to providing food pollinators.

Annual Plants

Annual plants are usually the first you will want to introduce to any flower bed or natural landscape, especially if you are growing vegetables, seeing as they are the perfect companion plants to protect your vegetables. They are easily identified by completing their whole life cycle in less than a year. This means that they grow from a seed to a full plant that produces its own seeds and die, all in less than 12 months. Other than tropics and sub-tropics, the majority of common vegetables are annual plants. Some of the popular annual flower choices include marigolds and petunias. Come rain or shine; you will see progress in growth during the same year with annual plants; they are quick-growing plants with blossoming flowers, which are great for beginners because they get to see their effort paying off without having to wait patiently for their plants to grow over the course of long months and years. However, one thing is worth noting; annual plants require twice the effort because you will have to harvest them and plant new seeds every spring. This can also be done through transplanting young plants.

Pros:

- Annual plants are quick to blossom.

- They grow rather quickly.

- They can easily turn a bare land into an established homestead in less than a year.

- They are considered nature's way of covering bare land.

Cons:

- You will have to replant them every spring.

- They need great amounts of water.

- They require regular and consistent work.

Biennial Plants

While annual plants take a year to grow and die, biennial plants need two years to complete their life cycle. During the first year, they blossom into flowers without producing any seeds. However, during their comeback in the spring the following year, seeds are produced. After that, the plants die off. Just like annual plants, there are some biennial weeds that you can use to cover bare land, such as mullein. That said, you should also bear in mind that there are some harsh conditions that can force biennial plants to act like annual plants. Take carrots, for example. They can grow to full-blown plants that can be harvested during the first year. If leftover winter, they produce seeds, but the plant itself becomes too rooted in the ground and tough to harvest. This is why many people prefer leaving only a few plants to overwinter to save the seeds without having to exert too much effort into harvesting all of them.

The problem with doing this is that when biennial plants grow as annual plants, they become less edible and more bitter once they start flowering. This may pose a problem for you if you are harvesting them for consumption. However, this can be more convenient if you're merely planting them for pollinating reasons. On the other hand, if you want perfect harvests, you will have to plant and harvest them every year to ensure that the taste is not compromised in the event that you plan to use them for herbal tea and concoctions. Ideally, you can cover your land with a heavy layer of mulch to prevent it from freezing. This way, you can use biennial plants for low-maintenance winter crops instead of having to store them in certain conditions to stop them from going bad. Common biennial plants that are perfect for this scenario are hollyhocks, brussels sprouts, beets, and kale.

Pros:

- Not only do many types provide a perfect harvest quickly, but they also supply a second year of blooms.

- They help in establishing bare lands and improving them without much effort.

Cons:

- They are replanted every year for the perfect harvest.

- They require great amounts of water.

- Second-year blooms might not be of the same quality as their first-year harvests.

Perennial Plants

Annual and biennial plants offer you some flexibility and freedom to replace your plants according to your needs since they come with relatively short life cycles that last for a year or two. With perennial plants, on the other hand, you have less flexibility since they will grow on your land for longer periods. However, this offers some peace of mind. You only have to plant them once, and they will keep coming back year after year without you having to worry about them. While they might require more time to grow into full-blown plants, they still require less effort from your part, and they save you time over the years. Take asparagus, for example. These are the most famous perennial plants that require at least 3 years after planting them before they can be harvested. The same applies to perennial fruits and berry bushes that may produce small harvests during the first and second years. However, perfect harvests only happen after the plant is fully established. Since perennial plants have larger roots, they require fewer amounts of water, which is perfect for your homestead at the early stages.

If you want to mix and match different types of plants in your homestead, you must be aware of how much room perennial plants grow to take. At first, if you plant the above mentioned three types, you should expect biennial and annual plants to grow more abundantly with their weeds and flowerbeds. However, after a while, perennial plants tend to become the dominant type of plants as they grow much larger than the former two types.

Pros:

- They don't require frequent watering as much as a biennial and annual plants.

- They save water.

- They continue to provide good blooms and harvests every year without having to replant them.

- They supply larger harvests once they fully grow.

- They require much less effort.

- They are perfect pollinators as they attract beneficial insects, bees, and beautiful butterflies.

Cons:

- Since you will not have the option to replant every year, you will have much less flexibility.

- To be able to benefit from perennial plants, you need to make some changes in your diet and cooking habits.

- They tend to take up larger spaces in your land.

The Ideal Setup

After you have familiarized yourself with the three main plant categories, you can now easily proceed to plan the ideal setup for your homestead. For starters, you should choose perennial plants as the foundation of your flower beds since they take much larger spaces and need some time to grow. Make sure to choose different types of perennial plants that are edible, pollinator-friendly, and come with different sizes and colors. After that, you can grow your own preferred set of annual and biennial plants that give you different beneficial additions to your homestead according to your needs. This way, you will be able to create a dynamic and vibrant habitat that will not only add a touch of color to your homestead, but they will also give you an abundance of different fruits, vegetables, and flowers that you can enjoy.

Introducing Pollinator-Friendly Plants

After you have planned the ideal setup for your homestead, you must consider introducing some flowering plants as pollinators. There are two approaches to this mission; some people prefer filling every nook and cranny in their garden with pollinator-friendly plants that also act as companion plants. Others prefer creating designated areas for pollinators to be able to have more control over their land. Deciding between these two styles of incorporating pollinator-friendly plants doesn't only depend on your preference, but rather it's more dictated by how much space you have, the type of plants you want to go for, and the weather conditions in your area.

There are a few rules to follow when you are trying to incorporate pollinator-friendly plants to your homestead. First of all, you must completely understand that the few examples included might not be completely suitable for the weather conditions in your area.

Therefore, you should only choose a variety that is indigenous to your land because they are already accustomed to the climate and can easily attract local pollinators. Secondly, including a variety of different plants that are well-suited to your climate at different times will maintain your homestead as a continuous source of food for pollinators. Mixing and matching between flowers that bloom at different times can also be enhanced by following succession planting, where you choose different times for your sporadic planting to ensure that you have food for pollinators all year long.

Thirdly, while incorporating different plants into your homestead garden, make sure to include plants with different heights, colors, and seeds to attract different pollinators. You can also add to your mix of flowers a few plants that attract butterfly larvae and act as their hosts, such as milkweed, fennel, and dill. Last but not least, it's essential that you learn how to practice organic gardening without the use of pesticides, especially when you are trying to attract pollinators. This way, you will create a healthy natural environment. Establishing a dynamic ecosystem will take care of pests and keep them in check by the use of beneficial insects and other helpful wild birds without any effort from your part.

Plants For Pollinators

As explained, when trying to establish a successful dynamic homestead, it's essential to choose plants that are well-suited for your climate to ensure success. Here are some examples of pollinator-friendly plants that will not only add beautiful hues of colors to your homestead, but they will also help you in building a healthy ecosystem that doesn't require much effort from your part.

Calendula Officinalis

These annual plants are the perfect choice for adding a colorful short bush to your homestead. They come in different colors, mostly with yellowish-orange shades that will provide not only pollen but also nectar for your pollinator friends. People who believe in alternative medicine can also benefit highly from this plant's medicinal healing properties. Calendula officinalis are pretty much daisy-like flowers and are technically used as herbs that can be added to your salad for a uniquely delicious taste. You can use them as companion plants to your vegetables in order to repel pests and attract beneficial insects such as ladybugs and hoverflies. They tend to bloom from the start of spring through late fall and can easily adapt to different soil conditions; however, partial shade in a hot climate is better for speedy growth.

Marigold

Just like calendula, marigolds are annual plants that act as companion plants with their beautiful shades of yellow and warmer tones of orange and red. They repel harmful insects such as cabbage moth. They grow taller than calendula, starting from only 6 inches and can grow up to 4 feet tall. Their height might start causing a problem for other plants, so pruning might be required to make them more suitable for their surroundings. Since they continuously bloom from late spring all the way to the end of winter, you can ensure that your homestead will attract butterflies all year long.

Lavender

We all know the endless benefits lavender holds as a plant. With its soothing properties and beautiful colors, lavender isn't only our personal favorite, but it also works great in attracting bees and repelling mosquitoes and flies. Lavender sizes and shapes vary, but they are usually tall spikes, which can be found blooming over small bushes that tend to have more silvery shades. Lavender acts as a perennial plant in some zones and an annual plant in others, but they bloom almost all year long, from summer throughout the fall.

Sunflowers

Way back in the day, this widely beloved annual plant has been recognized as one of the most beneficial and beautiful flowers in history. It has been domesticated since 1000 BC. People saw the value in planting and harvesting sunflowers seeds, not only for making oil, but dried seeds are the perfect snack for our feathered friends as well as for us. Sunflowers come in all sorts of sizes, shapes, and different colors. They have a wide structure that makes it much easier to attract bees. One thing that must be noted is that sunflowers must grow in an area where they can receive adequate sunlight as they tend to move constantly during the day to be in direct contact with the heat. Sunflowers are the perfect plant choice for incorporating a variety of heights in your homestead as they tend to grow taller than most flowers to the extent that they might require some support by staking.

Zinnia

The majority of homesteaders are familiar with the benefits of adding these vibrant, colorful balls of flowers to gardens. With shades that range from pink and yellow to lime green; and heights that range from short to several feet high, they are versatile and are known for being perfectly cut flowers. However, they are better left as delicious bait for our pollinator friends, especially monarch butterflies and other beneficial insects. While planting zinnia flowers, you must be careful to plant them where you want them to be as they don't tolerate transplanting very well. However, if you want to grow them inside and transplant them afterward, make sure to do so while they are not fully grown and before they harden. Once they are rooted in your ground, they become direct-sow plants that shouldn't be moved. Zinnias require a much more controlled environment with very specific requirements when it comes to their soil as they need rich compost soil to grow and turn into established plants.

Cosmos

These whimsical daisy-like flowers are a beautiful addition to any homestead with their various benefits and use. Different shades can be found that range from pinks and purples to more rare shades of chocolate brown, red, and yellow. Cosmos attract all forms of wildlife to your backyard homesteads such as bees, butterflies, birds, and other beneficial insects. You can make the most out of their petals and add them to salads or drinks as a garnish. Not only are they relatively easy to manage as you don't have to worry about any pest issues, but Cosmo plants can also be used as cut flowers. All you need is to scatter their seeds during the spring on bare soil after ensuring that the dangers of frost have passed. With little to no preparations, the cosmos can easily become fully grown plants in different soil conditions, and without the need for fertilizers.

There are no right and wrong species of flowers to plant in your garden; it all depends on your homesteading needs and whether you'll be planting herbs to harvest them for consumption, or looking to grow the perfect species that encourage pollination. However, in general, whatever you choose to grow in your garden will certainly prove useful to both your homestead and the environment.

Chapter 4

Designing a Homestead Garden

Homestead gardening is more of a state of mind than anything else. When you are brainstorming for design ideas for a backyard homestead garden, you should really be thinking about self-sufficiency as a way of existing. Granted, it can be quite challenging at first to figure out where to start when it comes to designing a garden, especially if you do not necessarily have enough experience in the area altogether. However, once you put your mind to being self-sufficient and applying that thought into everything you do with your garden, the rest will probably flow very easily. When it comes to designing any garden, understanding the basics is key to success. Here are some tips that can inspire you to utilize every inch of space you have in your backyard to plant the perfect homestead garden.

Measure Garden Space

The first step in making any garden design work measures the exact space you have for your desired homestead garden. You must measure every inch of space you have that can be utilized for growing that garden. Next, work slowly towards using that space effectively.

You would need to visually map out where your chosen plants would go in the homestead area and take into consideration that plants vary in size according to their types and as they grow. Make sure that your measurements for design allow for plants to grow in their own space without being crowded by other trees or plants that can later affect the overall layout of your garden. Use the right set of tools and equipment to take exact measurements and do not underestimate the importance of mapping out your garden as a first step as it can be a total game-changer.

Small Backyard Essentials

Size does not matter when it comes to planting a homestead garden. Whether you have a small suburban backyard or a wide space, you can always plan out a design that suits your needs and fits into the available space you have for self-sustainability. If you have a small backyard, then designing a homestead garden would have to be limited to the essentials rather than extravagant additions. You can start out by planting some fruit trees in a triangular corner, then add a herbs bed to another side, and maybe finish off with a small shed where you can raise poultry. The more space you have, the more essential trees of fruits and vegetables you can plant. Make sure that you start small and build your way up so that you do not end up overwhelming your soil or even yourself with all the required labor it takes to care for many plants in a small backyard. Small, homestead garden designs depend greatly on the smart usage of space. So, remember to make use of every little inch and divide the space into small, carefully engineered, and outlined spaces where you can plant one or two plants of everything essential.

Wide Acre Layout Plan

If you are blessed enough to have some wide acre space for a homestead garden, then designing a homestead garden that has all you would need to be self-sustainable can be pretty easy. You can add more crops and bring in more animals as wide as your acre space gets so that you can enjoy more sustainability. Start out with vegetable and herb boxes that are set up side by side, and add more of them depending on the kind of space you have. You can then include bigger fruit trees to surround your garden. Large trees would work as a fence for your private homestead garden space as well as a sustainable source of food. That said, wide acre layout plans would provide shed spaces for chickens and goats or any kind of animal that can provide meat and milk. Make sure you leave some space in your design to park any large equipment or vehicles, including tractors or other heavy machinery.

Business Homestead Designs

If your garden space is large enough to have a fully functional homestead garden with all the essential plants and garden equipment, then you could consider opting for a business design. That way, while being self-sustaining, you could also earn some extra cash. The only difference in design that you would need to make is making sure you leave more room for crops and animals that can make you more profit. Of course, that does not mean neglecting essential space for corps that you would need to use on a regular basis. However, you would need to widen the space for high demand crops and make sure your animal enclosures or sheds are built with business benefits on your mind so that you do not lose track of what your homestead garden is made for.

Edible Garden

All plants help in sustainability. But when you are planning on planting your own homestead garden, you might need to focus on edible crops that would cater to your needs more than anything else. Some people focus more on the looks of the homestead garden rather than the crops, and that is perfectly fine. However, if you are looking for more function than glamour, then designing an edible garden might be the way to go. The key to designing an edible garden is knowing exactly what plants can be planted side by side, and what kinds of trees would take up limited space but provide high crop results at the beginning of the season. You would have to make the most out of your space by literally putting plant beds at every corner and leaving room for planting herbs or small plants around larger trees. The more crops you can plant, the better.

Micro Designs

If you are looking to have the best of both worlds, that bring functionality as well as a gorgeous layout, then a micro design for a homestead garden might be what you need to work on. Unlike many other designs for homestead gardens, a micro design would not provide much space for animals. However, you would be able to plant more fruit and vegetable trees, as well as ornamental plants for a beautifying element. Ornamental plants would not necessarily fall into the self-sustainability theme, but you have got to remember that planting any tree, big or small, still counts as a sustainable contribution. It would also offer an elegance element for your homestead design.

Plan for Crop Rotation

Plants are living beings that exist to serve a purpose and, in most cases, provide useful crops. However, when you are designing a homestead garden, it is essential to realize that crops will not simply keep on growing all-year-round. For some plants, once they produce their crop for the season, they would simply continue to exist as ornamental plants rather than fruit or vegetable-providing trees. Other plants would provide crops for a few years, then wilt after serving their purpose. This is why it is important to take into consideration when designing a homestead garden that crop rotation plays a crucial role in the functionality of your homestead. There may come a time when the trees and plants would have to be replaced.

All-Inclusive Designs

Having an all-inclusive homestead garden design is all about implementing self-sustainability strategies in every little inch of the garden and the surrounding buildings on the property. It does not matter if the space is large or small, you can have an all-inclusive design with any size space if you plan it smartly. This means that you might have to include energy sources, irrigation plans, plant beds, and maybe even animal sheds in the garden design. It would be even better if you can make the entire garden green in terms of the energy sources used, and limit any waste by implementing smart designs to include all you would need for self-sufficiency.

Big Family/Small Family Layout

For many people, having a homestead garden comes in handy for practicing a healthy hobby that has numerous rewarding outcomes on the person caring for the garden, as well as the environment in general. However, some people look to have a homestead garden as a way of living. If that is the case for you or for the person you are creating a design for, then you should consider the family needs in the design. If the family is big and they have enough acre space, then you should make more space for fruit trees and animals to live in the farm garden. If the family has few members or if they have a small garden space, then take a look at the essentials they would need and design an effective garden that would allow the family to live comfortably and have all the plants and meat they would require. Make sure that you put into consideration the living space of the family residing by the homestead garden so that you can design the space that allows them to cultivate useful crops, as well as enjoy a cozy living space.

Design Irrigation Paths

One of the most important things to consider when designing a garden is irrigation paths for watering your plants and trees. Many homestead garden designers get too distracted when planning a layout for the garden and forget about the need for daily watering. This plays an important role in the success of your endeavor, so this step should never be taken lightly. Irrigation paths are among the first factors to consider in your garden plan, maybe even before you consider where the plants would go, as the location of certain plants could heavily rely on how easy it would be to provide them with sufficient water. When you think of water for your homestead garden, you would have to factor in any animals you may be raising as well. This could mean designing water tanks or large pots for them to stay hydrated and healthy, especially on warm days.

Making Room to Enjoy the Garden

Homestead gardens are not just about having enough food to sustain yourself. They can also be great areas of green space to enjoy yourself and have a relaxing time at your own property. When designing a homestead garden, you should remember to create some paths or hangout areas where you can sit or have walks, whether alone or with loved ones, to enjoy yourself and have some quality time. If you have enough space, you can even create large spaces where you can connect with nature by simply sitting with your friends and family right by the plants or doing fun activities.

Considering Weather Conditions in Design

Weather conditions affect the state of any kind of garden immensely; most of all, a homestead garden planted to provide essential crops for self-sustainability. When designing the garden, you should be mindful of the kind of weather conditions the area where the garden is located is known to have. If the weather conditions are too extreme, then it might be wise to design greenhouses for some of the plants to grow safely. Different plants would need different design arrangements according to their compatibility with one another and by weather conditions. Make sure that you do some research or consult with a gardening professional before you finalize the homestead garden layout plan.

Designing the Garden-Based on Soil Conditions

Similar to weather conditions, soil conditions affect different types of plants and trees. Not all plants can be grown in just any kind of soil, and that is something that you would need to keep in mind when designing a homestead garden. Again, you would need to do some research on what the plants you have in mind need specifically and work your way around it when designing the layout. If the soil does not match what the plant needs, consider adding some high beds or pots where compost can be placed with fertilizers to ensure that the plants grow effectively.

Consider the Digging Process

Designing a homestead garden is not just about drawing a plan on paper; it is about implementing that plan in real life in the space of your backyard. One of the main factors to consider in your garden design is the digging process, which you will need to educate yourself on before proceeding to plant trees and place seeds in the soil. There are some design considerations to make ahead of the digging process, including whether your soil is sufficient for the process, or if you'll need to add new soil on top of the existing one. Some homesteaders opt for only pots and beds, rather than planting in the ground soil itself if the soil is not suitable. It all comes down to the different factors that can affect your decision when designing.

Rows and Wide Rows

When it comes to planning the actual layout for your homestead garden, you would have several options to place your plants or any trees in an organized manner. Rows and wide rows are among the most popular ways of setting out a neat garden layout that allows for good air circulation and opens the space for irrigation paths and walking lanes. Wide rows, in particular, are great for gardens planted in small backyards, as they allow for a large number of plants to be planted next to each other, and still look elegant and fresh. The row design in homestead gardens also has the perk of keeping the soil moist for long periods of time, to avoid spending fortunes on watering the plants too much.

Check Catalogues for Inspiration

There are tons of different design ideas out there for homestead gardens. Depending on the kind of space you have and the kind of plants you want to have in your garden, the design can differ massively. Many people struggle to settle on one design or on implementing a certain design that they have in mind for the kind of space they already have. Checking catalogs for inspiration can help with that process, as it can offer real-life examples that can get your creative juices running. Design catalogs can help you make useful additions to create your ideal homestead garden and see all the new and cool layout plans that can be incorporated into your backyard space.

Being self-sufficient is something that many people nowadays are keen on. By designing your own effective homestead garden, you can do just that! Homestead gardens are all about living off what you make. By planting the right crops and raising enough animals, you can ensure that you have a sufficient amount of food that you would need for a lifetime. All it takes is some research and making professional consultations with gardeners, as well as making smart decisions about the usage of space, and your garden would be ready to blossom. Remember to take water, soil, and weather conditions into account in your design to make sure that the plants and animals in your sustainable garden remain healthy and offer you a worthwhile small farm experience.

Chapter 5

Care and Maintenance

Gardening is a wonderful experience, especially to homeowners who are concerned about improving the appearance of their homes. The biggest perk about gardening is that you can do it yourself as long as you have the right equipment to use and knowledge about how you can perform the different tasks required for you to handle patiently. While many of us may not have the time to attend to our outdoor spaces to make them accommodating, small details can make a big difference. As such, this chapter highlights some garden care and maintenance tips that you can consider while improving the appearance of your garden. There are different things that you need to consider so that you can create an ideal outdoor space that can improve the value of your homesteading area.

Location Of Your Garden

Do you know that the location of your garden plays a critical role in determining the growth and maintenance of different plants? If you choose the backyard as the primary location for your garden, it might not achieve much in terms of improving the appearance of your home unless it is strictly utilized for growing vegetables. All the same, a vegetable garden thrives in a location with access to abundant sunlight. It is also important to ensure that you locate your flower and lawn garden in a strategic position that can improve the appearance of the entire home. This also helps create a lasting impression on your guests.

Lawn Maintenance and Care

Watering is an important aspect of lawn maintenance, and it is crucial to keeping it evenly moist. Depending on the climate and weather conditions in your region, it is essential to water your lawn at least one to two times a week. Another important thing that you need to bear in mind is that you should water your garden in the morning or evening so that less water will evaporate. While watering your garden is beneficial, you should know that many diseases also thrive in humid conditions and require water for growth, just like plants. For instance, pathogens in the soil require water to reproduce, grow and move, which spreads the disease to different places in your garden.

For that reason, you should choose watering methods that will limit the amount of moisture on the foliage. You can accomplish this by choosing drip irrigation or soaker hoses to water your garden. If you are watering your garden by hand, make sure you prevent water from getting onto the leaves or plant heads since this can lead to mold formation. Watering directly on the plants also leads to damage, and you should avoid it.

Another important tip that you should know when watering your garden is that more is not always better. Excessive water can lead to waterlogged soil, which, in turn, affects the growth of your plants. Waterlogging suffocates the plant roots, which can lead to decay and can also promote the development of unwanted diseases.

Irrigation System

If you are always away from home, this will not get in the way of growing and maintaining a healthy lawn. You can consider a computerized irrigation system for your garden that can regulate the amount of water your lawn gets when you are not around. The ideal time to water your garden, as you are now aware, is in the morning before 8 am or in the evening after 4 pm. You also need to protect your lawn against excessive heat, and you can do this by avoiding cutting the grass when it is too hot.

Prevent Growth Of Weeds

To maintain your lawn in good health, it is essential to prevent the growth of foreign weeds, which can also make your garden unsightly. Lawn weeds come in different forms, and they can affect the quality of your garden. To prevent weeds from sprouting in your garden, you can make use of mulch, but you need to be careful in the process, as you will see later that it can also affect aeration of the soil. Make sure that there are no patches on your lawn if you want to prevent the growth of foreign grass.

You can also remove the foreign weeds using a special tool like a weeding trowel. Weeds can be brought into your garden by pets, other animals, or can be blown by the wind. To rid the garden of weeds, it is important to pull the whole weed together with its roots. You can do this by hand to ensure that all areas in your garden are free of unwanted grass. Alternatively, you can use a herbicide to kill the weed, but make sure that you know different organic herbicides that you can use without harming your plants or the surrounding bees if you're also investing in a beehive.

Improve Drainage

It is essential to improve the drainage system in your garden to encourage the healthy growth of your lawn. When your lawn has poor drainage, it becomes waterlogged for several hours or even days, which negatively affects its growth. This will lead to stunted or poor growth of your lawn, so you should ensure that your garden has proper landscaping to promote the free movement of water. Waterlogging can also be caused by your soil's inability to absorb water, and this mainly depends on the type of soil you have. For instance, soil compaction, thick layers of thatch, and a high level of clay can contribute to waterlogging. These factors prevent water absorption, and it can end up pooling on top of the soil with nowhere to go.

To improve soil permeability, aeration is another option that you can consider since it significantly lowers the build-up of water. When the soil is well aerated, water can easily flow. You can achieve this by adding manure, which contains loose particles that promote quality aeration. Organic materials found in manure help promote the decomposition process and substance breakdown, which helps in loosening the soil to promote aeration. Another method that you can consider to improve aeration of the soil is to create small holes at certain depths and intervals around your garden. You can use your garden fork or another specially designed tool to aerate the soil.

That said, you should also consider shaping your garden in such a way that it naturally drains water away from your house. For instance, you can maintain a slope that promotes the flow of water from your garden to the main drainage system to avoid challenges such as waterlogging. It is essential to adjust the gradient of your garden landscape so that it can direct excess water to the main drain. You can also consider adding different plants that can thrive in excess water.

Mowing and Edging

Yet another critical aspect is frequently mowing your lawn as needed to ensure that you keep it in good condition, both aesthetically and functionally speaking. When mowing your lawn, you should use the right mower that is suitable for the size of your garden so that it can remove a sufficient length of the grass. How frequently you mow your lawn depends on different factors, such as weather conditions as well as the time of the year. During the winter, you won't mow your lawn so much, if at all, depending on weather conditions. In contrast, the frequency of summer mowing is increased given that warmer weather promotes quicker growth of grass.

There are certain steps that you should take when mowing grass in your garden. It is important to avoid mowing when the grass and soil are wet. This will cause some damage to the grass, and it can also affect the healthy growth of the lawn in the future. After mowing your lawn, you can use a half-moon edging tool or edging shears to shape the lawn around the edges where the lawnmower cannot reach. You can shape your lawn the way you like to ensure that it improves the aesthetic appearance of your yard. You can also use the shears to trim the shrubs and other plants in your garden to improve its overall appearance.

Use Appropriate Fertilizer

It is important to use the appropriate fertilizer on your lawn or any other plants in your garden. The problem with fertilizer is that it may contain some chemicals that can burn the roots if overused. This affects the plant's ability to absorb water and can cause the plants to wilt and eventually die. If the plants are weakened by fertilizer, they become susceptible to heat or cold weather. Small plants are usually affected by excessive fertilizer, so make sure you keep your eye on some red flags, including the spotting or yellowing of leaves. Too many nutrients can stress a plant, so if you want to stay on the safe side, you should ensure that you supply your garden with just the needed quantities of fertilizer and not more.

It is important to do a soil test every now and again to keep your plants from wilting. You can do this through your local extension agency. This helps you get accurate information about the quality of your soil and its nutrient level. The problem is that if you lack correct information about the quality of your soil, everything you're doing can merely be guesswork. This may not give you the desired results when you use fertilizer because you will either be undersupplying your garden with a particular nutrient or overdoing it, which can affect plant growth. Just like humans, plants should also get balanced nutrients to enhance quality growth.

You can also consider using organic manure or compost if you want to maintain a lush garden with minimal hassle. Compost consists of waste material that mainly comprises dead grass, leaves, and other biodegradable materials that can decompose to form manure and support plant growth. Composting can raise the temperature of your soil, which leads to the decomposition of the waste material to form manure. The advantage of using compost in your garden is that it is free of chemicals since it consists of all-natural nutrients. There is no fear of oversupplying or undersupplying the plants, as long as they are provided with sufficient water.

Mulching

Before you proceed to use the compost at hand in your garden, you need to know that you must avoid using just any form of waste that you find in your yard, namely mulch, on sensitive plants. You can never know if the debris is not infected with the disease, and you would only regret later after infecting your healthy plants. The buildup of debris can create a blockage in the soil that ultimately prevents essential nutrients and moisture from penetrating the soil to reach the roots. This can affect the growth of grass, as well as other plants that you're growing in your homestead garden. Thankfully, you can easily identify the affected places, since they consist of dead patches on the lawn and the ground will often have a spongy feel.

To prevent this problem, it is essential to remove thatch using a process called scarification, which involves the removal of excess mulch from the grass. You can use a rack or a special lawn scarifier to remove mulch from the lawn and to promote aeration of the soil and free movement of the water. Additionally, excess mulch on your lawn can also lead to diseases that will ultimately affect its health.

However, when maintaining a vegetable garden, you should add mulch around your plants to help keep the soil cool. Mulch also helps retain the water in the soil and will gradually compost, thereby adding nutrients to the soil to improve its quality. Mulch is a good fertilizer for a vegetable garden; however, you need to be careful when you go about this process and ensure that the mulch does not import foreign elements that can affect your plants. You also need to check if the mulch shows signs of disease before supplying it to your garden. As you are going to see in the following section, this is a crucial step to preventing disease from infecting your plants.

Plants and Disease Control

One of the most frustrating things in your garden is to see your plants succumbing to the disease. You will have more questions than answers trying to solve the puzzle of how the disease got there in the first place. You may also be fretting about whether the disease can be treated or controlled. However, the correct way to go about this is to prevent disease to ensure that your garden is safe; prevention is much easier than cure. Plant disease can be caused by fungus, virus, or bacterium, and it can also be developed due to environmental conditions such as drought or humidity. If none of these conditions are present, you won't have to worry about infection. It is, therefore, important to prevent disease by ensuring that no condition that can lead to the problem existing in the first place. Do not wait for the problem to appear in your garden before acting appropriately. It is crucial to be proactive. You can take different steps to prevent the disease from harming your garden. Here's how you can do that.

How To Prevent Disease

You need to keep a close eye on your garden and make sure that it is free of foreign elements that can bring about disease. It is a good idea to read magazines about gardening so that you can get a detailed insight into what can affect plant health. However, as a general rule of thumb, you should never take a plant with dead spots or aphids into your garden since it can be the source of infection that can later spread to the entire garden.

Once the disease spreads, it becomes increasingly difficult to get rid of it, so you must avoid it in the first place. You should also inspect the root of the plant to check if it is healthy before you plant it into your garden. The roots of healthy plants should be white and firm. On the other hand, if the roots are dark, they can rot at any time and thus negatively affect the rest of your garden. You need to check on all the plants you intend to plant before placing them in your soil, seeing as this can also affect your soil's health.

Clean Up

The second step that you should take to prevent disease is to clean up your garden regularly, especially during the fall. This helps you to control any disease that may already exist in your garden while at the same time, preventing conditions that can lead to the development of infection. Dead leaves and other debris can affect the new plant leaves during the springtime so you should ensure that your garden is clear of them. For instance, diseases such as black spots on roses, Iris leaf spots, as well as daylily leaf streak are some of the diseases that can be avoided if your garden is clear of dead leaves.

Prune the Plants

Trimming trees and other plants in your garden helps maintain it in good shape while at the same time works to prevents the spread of disease among your plants. You can prune your plants during late winter so that any kind of infection does not easily spread to new growth. When trimming your plants, it is essential to use sharp tools that do not damage the plant tissue, which can make it vulnerable to infection. When pruning your plants, make sure you do it when the weather is favorable so that they can heal quickly.

Additionally, you should know that pruning is as crucial as harvesting. You need to prune perennial plants in your garden during their dormant season. Such plants include fruits, and as such, you should get rid of old branches without damaging the plant. This will help new shoots to develop that will bear fruit. Another advantage of pruning your plants is that it will keep them healthy, and they can go a long way in producing more fruits.

Space Your Plants

You should ensure that other plants in your garden, apart from the lawn, are well spaced to promote healthy growth, while at the same time preventing the spread of disease. When plants are crowded, they create excess humidity, which promotes infection through rust, downy mildew, and powdery mildew. Foliage can dry quickly when there is sufficient airflow. Additionally, crowded plants also compete for water, light, and nutrients, which can affect growth. You should also bear in mind that weak plants are susceptible to infection more so than your healthier crops so you should make sure that they get sufficient nutrients that can promote healthy growth.

Support Your Plants

Certain plants in your vegetable garden may require support, as they will be heavy as a result of bearing offspring. For instance, tomatoes require a stake placed in the ground together with a string to tie them, since they may not be able to support the heavy load. Failure to support the plant can result in damage, and it can also lead to the products getting infected if the plant is left lying on the ground. Supporting your plants can also help you enjoy free movement in your garden.

Include Hard Landscaping

Apart from these care and maintenance tips, you can also consider hard landscaping around your home to improve its appearance and value. For example, hard landscaping that includes pebbles, paving, or decking helps improve the drainage when it is raining. Water cannot easily pool on a hard surface with a sloping gradient that is designed to enhance the flow of water away from your yard. Hard landscaping can also protect your property from water damage. That said, paving helps mark the sidewalks so that people will not disturb the plants in your garden.

Garden care and maintenance help improve the appearance of your home. There are different steps that you can take to maintain your garden, and luckily, you can do it yourself as long as you have the right tools and knowledge. Lawn maintenance is one major thing that you should focus on since it covers the greater part of your garden. To succeed in garden maintenance, you need to keep it free of disease and also make sure that you follow the needed steps to keep the soil in good condition to support lawn growth.

Chapter 6

Tilling and Planting

A midst the noise of the fast-paced life that we are currently living in, many people have taken a stand and decided to challenge the status quo. Hoping to break free from the sometimes choking grips of civilization, these people are looking to go back to the basics. Homesteading has become a celebrated trend among those who believe in the importance of self-sufficiency, from growing your own food to caring for your own land. In the previous chapters, we talked about how to design and set up your own garden and how you can select the ideal types of plants to grow. In this chapter, we will get more technical and discuss the different methods of tilling and planting that you will need to learn about to maintain your homestead. Finally, we will close this chapter with some flower suggestions that you can easily grow in your backyard. Let's get started.

What is Tilling?

If you are new to this world and still learning about the basics of gardening, you may not be familiar with tilling. Tilling refers to turning your soil to aerate it and mix the organic matters in to keep it healthy; it will also enhance the quality of your crops. However, tilling is not a random process; there are an optimum depth and frequency for tilling; otherwise, you can damage your soil altogether. According to experienced farmers, the best way to find out whether your soil is ready for tilling or not is to test it yourself. Scoop a handful of soil and make a ball, then poke it gently with your finger. If it smoothly crumbles, this is a clear sign that you can start tilling. Here are some of the most popular tilling methods for homesteading that you need to learn about.

Hand Tilling

For a backyard garden, hand tilling can be more than enough to work your soil before planting. Using a shovel to double-dig, you can start by spreading compost all over your soil, then start digging ditches and moving the shovelfuls around from one ditch to the next. This method is simple and doesn't need intricate planning. You can even have your kids help you out, they will have fun, and it will be a good way to teach them about the important basics of gardening.

Rototilling

If you have a bigger piece of land that would be challenging to hand till rototilling might be a good idea. Using either a motorized or a push rototiller, you can rotate your soil and break down any clustering to make it planting-friendly. However, if your soil is full of sods and tangling weeds, you might want to leave the rototilling job to an experienced farmer to get the best results.

No-Till Methods

No-till methods had become popular recently when farmers realized that soil doesn't need to be disrupted with tillage to maintain its quality. Instead, using either conventional or organic no-tilling methods, you can de-weed your soil and get rid of any harmful substances trapped within. Conventional no-tilling depends on herbicides to get the job done. However, if you prefer not to use any harsh chemicals around your crops, organic no-tilling might be a better option. With organic no-tilling, you use a special cover to kill off the weeds before you can rotate the soil to plant your crops. No-tilling is the least invasive method that you can use to get the benefits of tilling without compromising your soil. As a beginner, it might take you a few trials to get it right, so you might want to consult a more experienced farmer to show you how you can manage it on your own.

Shallow Tilling

Shallow tilling is an in-between method that provides the best of both worlds of conventional and no-tilling methods. Here, the aim is to disrupt the soil as little as possible while still reaping the benefits of tillage. Using few to no equipment makes shallow tilling gentle enough for your soil and helps preserve its fertile nature. If you have a big enough backyard to raise draft horses, you can use horse-drawn equipment to perform shallow tilling. Granted, you will have more work to do the feeding and looking after the horses; however, it is one of the cleanest and environmentally-friendly methods to till your land. Another way to shallow till is to use a chisel plow. A chisel plow is a simple piece of equipment that has double-ended shovels used to de-clump the soil and loosen it in preparation for planting. Depending on the size and nature of your soil, you will decide which shallow tilling method is better and easier for you to manage.

Regardless of the tilling method that you end up using, you have to understand that it will take some time before you can master the process. Now that you have your soil free of weeds, well pulverized, and have all the good organic matter filling your soil to the surface, it's time to start planting. The main reason you want to build your own backyard homestead is probably to ensure that your family has access to organic, high-quality food. So, make sure you keep that in mind when selecting one of the many planting methods that we will discuss throughout the rest of this chapter.

Where to Start?

It's not only a matter of what you want to grow. You have to understand what the crops that you can grow are. Deciding on the plants and flowers that you will be growing is a good place to start. Create a list of the fresh produce that you use the most, then reach out to one of the local farmers to discuss the possibility of growing those in your backyard garden. Local farmers have a lot of knowledge regarding the nature of the soil where you live and the climate, so they can advise on the crops that have the best chances of growing and thriving. During this phase, try to keep an open mind and consider growing produce that you can use in many different ways, even if you aren't particularly familiar with it. When you are first starting, it's a good idea to choose crops like lettuce and cucumbers, which are easy to grow and give fast results, encouraging you to keep going. Next, it will be time for you to buy the seeds. Browse online to find where you can buy the best quality seeds in your area, as this can make a world of difference in how your crop turns out.

Different Planting Methods

Even though the size of your backyard can limit your options, you still have a number of planting methods to choose from. Below we will discuss the most popular kinds.

Intensive Gardening

Intensive garden refers to reducing wasted space of land to a minimum. In other words, it means filling your backyard with as many crops as you can possibly manage. Not only will this method allow you to grow all the produce you dream of, but it will also simplify your gardening chores and make them more manageable. It's like hitting two birds with one stone: less work and more yield.

Interval Planting

If you want to make the most out of your garden, interval planting is a smart planting method that you can benefit from. Using the same land to grow winter produce, and follow it by summer produce when the season's change, is one example of interval or succession planting as it is sometimes called. Even with similar crops, interval planting is sometimes the better option so that you can ensure a constant smaller supply that you can consume per your needs. Otherwise, if you plant everything all at once, you might end up with more produce than you need and end up letting it go to waste. If you are into experimenting with different varieties of the same crop, using interval planting, you can plan to have an early, late, and primary seasonal harvest. Interval planting requires proper knowledge of the crops you are working with, so it might be a good idea to work with a farmer to avoid messing up your crops for the season.

Raised Bed Gardening

Raised bed gardening is a good way to divide up your small backyard to grow your produce with good drainage and away from pesky pests. Using bottomless gardening boxes that open directly into your soil means that you can still grow your favorite root vegetables without having to worry about them having enough space to grow. Another huge benefit of raised bed gardening is that it will make it easier for you to carry out your chores and de-weed your crops without straining your back for long hours. If you choose to use raised beds, you will also have the advantage of starting planting early before the actual season since the soil in the elevated garden boxes is warmer and better drained.

Vertical Gardening

With vertical gardening, you can double the size of your land by hanging some plants that grow on sprawling vines using special poles and strings. This is an excellent way to increase your yield if you aren't ready to experiment with interval planting. Hanging your plants this way will usually make them dry up fast, so they will need more frequent watering. You also have to be mindful of the plants that you plan on growing underneath the hanging ones because they will have limited access to sunshine. You can dedicate this spot to arugula and spinach since they can grow and thrive despite lack of sunshine.

Cold Frame Gardening

If you live somewhere cold and wet, you need to understand all about cold frame gardening. The Cold Frame gardening method gives you the chance to enjoy warm-weather crops regardless of the season. Cold frames are basically see-through boxes that can lock in solar energy to isolate the plants from the outside cold weather and frost. To get the best results with cold frames, there are a few tips to keep in mind.

1. Find the best location to place the cold frames. You want to pick a spot that has the best exposure to sunlight while at the same time is sheltered away from strong winds.

2. Select a suitable material. According to where you live, you will decide which material can work best for your cold frames. If you live in an extremely cold climate, your best bet would be to choose cold frames made from bricks and polycarbonate to provide enough insulation for your budding plants.

3. Ventilate the cold frames properly. You can equip your cold frames with a hands-free vent opener that will automatically open the top and aerate the insides of the cold frame when the temperature rises.

4. Always clean the cold frame tops. Plants growing inside cold frames already have limited exposure to sunlight, so you must make sure that you don't have any dirt or debris covering the tops, which can stop what little sun they get from getting through.

5. Help your cold frames retain more heat for longer. You can use aluminum foil to cover the interiors of your cold frame to retain more heat for your plants to maintain their consistent growth, making sure not to cover the top.

Think About Aesthetics

Having your own backyard garden doesn't mean that you have to think only about practicality and growing enough food to eat. Considering the aesthetic aspect is very important when growing your garden. You have to make sure that it is pleasant to look at. Planting some flowers among your produce will add a beautifying touch and will allow you to, later on, do your own beekeeping, which we will discuss later on in the second part of this book. When it comes to flowers, as you would expect, the best time to start is during the spring. However, it all comes down to which flowers you are interested in growing. Below, you will find some suggestions for you to consider as well as specifics regarding each flower.

Calendula

This medicinal flower is a common ingredient in skin care products as it helps in fighting skin inflammations. Calendulas are beautiful to look at and are a favorite among garden pollinators. They also act as natural pest repellents, so it's a good idea to plant them among your produce for protection. In addition, you can use calendula in cooking as an aromatic herb, or even drink it as a tea if you suffer from digestive issues like acid reflux. It doesn't take much effort to grow calendulas. All you need is abundant sunshine and good quality seeds to start with. After your flowers bloom, it is encouraged to harvest them frequently, as this will keep them growing further.

Sunflowers

Not only do sunflowers they look pretty, but they are also incredibly inviting for bees if you are thinking about beekeeping in your backyard. This might come as a surprise to you, but not all sunflowers are created equal. Make sure you grow the kind that has pollen to support your beehive. If you plan to raise a chicken as well, end of season sunflowers is an excellent source of nutrients for your birds. As you probably know, sunflowers love to face the sun; in fact, the flower heads move throughout the day to make sure they are getting as much sunshine as possible. You won't need to water your sunflowers too often as they are known to be drought-tolerant that can suffice with little hydration. Remember to save some seeds aside for yourself that you can roast and enjoy as a healthy snack.

Marigolds

These bright yellow flowers, like sunflowers, play an important role in keeping harmful insects away from your crops. Furthermore, they kill off parasites that dwell deep within the soil and feed on the roots of your plants. Farmers recommend using a no-till method with marigolds to keep their roots intact. Marigolds are edible and can be used in a variety of dishes as a garnish, and they also have a number of health benefits. Marigolds grow better in warm weather where there isn't a shortage of sunlight. Keep their soil adequately watered, but make sure you don't overdo it as this may inhibit their growth.

Borage

The blue borage flower attracts bees from miles away. You can also use them in the kitchen, as they have a strong, distinctive taste. What is special about borages is that after they absorb nutrients from the soil, they can store them in their leaves. They can then be used as natural fertilizers to keep your garden organically lush all-year-round. Borages grow equally fine under the sun and in the shade. However, if you want bigger blooms, keep them away from the sun. These flowers are very resilient and can pretty much grow in any kind of soil.

General Planting Tips

It would be a shame to let your flower garden wither after you put in all the effort in tilling and planting. While each type of flower needs special treatment, there are some general tips that you can follow to maintain all of your flowers.

- Use the right tools. If you want to make your flower garden a success, you have to invest in some gardening tools early on. They don't have to be fancy but make sure they are made from high-quality material to last you for years to come. Start small and build your way up to more sophisticated equipment when your garden grows lusher.

- Perfect your soil. Your soil quality will make or break your whole gardening experience. Give it the time and attention it deserves until you get it to a satisfactory state that you can rely on to support the growth of your precious plants. Don't get discouraged if you fail a couple of times before you eventually get it right. With time, you will learn how to pinpoint any issues with your soil immediately so you can fix it. Make testing your soil a habit to make sure it's ready to go. Regular testing will also help you tell when the time comes to replace your soil. As much as you can, steer clear from chemical pesticides to keep your garden as green as could be.

- Keep a detailed schedule. It's very easy to get flustered and confused with the different dates and relevant chores when you are first starting. To make your life easier, draw up a schedule for each plant in your garden, highlighting all the

356

relevant dates. You can even use a mobile app to send you notifications before these dates come up.

- Prepare your own fertilizers. Your topnotch soil deserves topnotch fertilizers. Preparing your own fertilizers is the best way to guarantee its quality. Use compost and produce scraps to enrich your soil with beneficial nutrients. Preparing fertilizers is one of the healthiest and environmentally-friendly ways to get rid of food remains.

- Always remove weeds. Weeds can undermine the integrity of your soil and destroy your crops. Always make sure to remove weeds using a safe and effective method. Use a hoe and a pick every couple of weeks to destroy weeds in a natural way.

- Protect your flowers against diseases. Diseases caused by bacteria and viruses can travel through your garden fast if left untreated. Use clan weeding methods to protect your flowers from these harmful microorganisms. Also, be careful not to overwater your garden, as it might cause fungal infections.

- Keep your tools clean. Just because you're using your gardening tools in the dirt doesn't mean you don't have to clean them. In fact, you have to clean and disinfect your tools before every use. You can use a damp towel and an isopropyl alcohol spray to give your tools a good scrub.

As you can see, there are tons of details that go into preparing your soil and planting your flowers. Each plant needs special care to give you the best results. However, as you are building your backyard homestead, try to enjoy the process and revel in the fact that you are doing something completely new and building your own haven. As mentioned above, in the next part of this book, you will learn everything there is to know about beekeeping so you can build your own hive.

PART TWO

Beekeeping

Chapter 7

Parameters to Consider

Your intention to become a homesteader must have had something to do with your love of nature and your desire to befriend our planet. In the first part of this book, we talked about designing and setting up your flower garden. By now, you should have a clear idea about how to prepare your garden for its most valuable inhabitants - the bees. Bees are, without a doubt, one of the most important creatures in the food chain. We rely on these tiny creatures to pollinate almost one-third of the food that we humans consume on a daily basis. Unfortunately, though, our careless behavior and reckless actions have endangered bees and pushed them towards extinction. The good news is that there is still hope to save the bees if more people like yourself decide to build their own beehives and look after our buzzing friends. To start off this part of the book, we will be discussing the parameters you need to consider for beekeeping in this chapter. We will cover everything there is to know regarding cost, location, climate, and everything in between. But before we get into the meat of this topic, let's first take a look at the other less obvious reasons to keep bees.

Why Beekeeping?

Besides what we mentioned earlier about bees being the main pollinators, there is a multitude of other less-celebrated benefits to keeping bees. Here's how beekeeping can help your homesteading goals.

Maintaining Your Luscious Garden

Bees don't only suck the nectar from your blooming flowers to nourish and produce honey; they also have a magical effect on the yield of the trees they visit.

For the Honey

This, in and of itself, is a good enough reason to keep bees. The natural sweetener has a myriad of benefits and is known for its potency in healing many ailments.

Starting Your Own Business

If you are looking for a side hustle to make some extra cash, you can trade your quality honey in the local farmer's market, or if you wish, you can go bigger and build your own honey empire. The best thing about the honey business is that you won't have to worry about excess supply since it can never go bad.

Developing Your Kids' Infatuation with Nature

Watching the bees working tirelessly in the beehive is nothing short of a miracle. If you want your kids to have a better idea about why you uprooted them from the city into the suburbs to become homesteaders, a beehive will be a valuable prop that you can refer to.

Strengthening Your Commitment to Homesteading

When you are first starting your homestead, it is only natural to feel discouraged the moment you realize that there's a lot to learn. However, if you take a step back and contemplate your growing beehive, you will realize what you are contributing to the world. You will become even more adamant about being an advocate for the environment.

Now let's shift gears and get into the technicalities of beekeeping. Preparation is key in your journey to become a homestead beekeeper. If you want to build a sturdy beehive and ensure that you will have the best-quality honey, take it one step at a time and don't skip learning about any of the below essential parameters.

1. The Costs of Homestead Beekeeping

Besides the bee packages and the hive, you will need to invest in some equipment to get started. Here is a rough breakdown of the cost of each and some important factors to consider before purchasing.

- **The Hive:** There are many variations of honeybee hives that you can buy. However, according to experienced beekeepers, a Langstroth hive is the perfect choice for beginners. Langstroth hives are made of wooden boxes stacked on top of each other and removable frames that the bees can use to build combs. You can customize your Langstroth hive as you see fit. Starting with 3 medium boxes that are 60 pounds each is a good setup for you to begin with. Although the hive can be quite heavy, it's actually a good thing as it discourages you from moving it around, which is strongly advised against. A fully equipped Langstroth hive will cost you around 200 USD. This is, of course, a round figure as you have to factor in shipping costs and other special additions that you will need according to the number of layers you are planning on having. You might be tempted to look for second-hand options. Even though recycling and sustainability are important pillars in homesteading, when it comes to purchasing a hive, always buy new ones. Pre-inhabited hives can be carrying diseases and other harmful contaminants that can put your bees' lives at risk, or worse, kill them off as soon as you release them inside. The risk is far bigger than you can afford, so think of your hive as a one-time, long-term investment that can last you for years to come.

- **The Hive Equipment:** To furnish the hive and prepare it to receive the bees, you will need to do some painting and insulation to make sure your bees are protected from the ever-changing weather conditions. You will also have to buy beehive support to hold the hive and allow for aeration. These items aren't necessarily pricey; however, if you are planning on having more than one hive, they will definitely add up. That said, 20 USD per each hive is a fair enough figure to account for.

- **The Bee Packages:** After having set up the home, it's time for the inhabitants.

Bees intended for beekeeping are sold in a bee package, which is typically 3 pounds of bees and roughly 10,000 worker bees with an isolated queen, all sold in one box. It is important to release the bees from the package into the hive once you receive them as the packages are not meant for long-term use. Most popular honey bees sold in packages are the Italian and Carniolan honey bees. The two types differ mainly in the temperament of the bees, so for a beginner like yourself, it might be better to start with the Italian bees as they are easier to handle. However, both types will cost you around 200 USD to 350 USD per package. The growing demand for bees compared to the waning supply is the main reason behind the noticeable increase in bee packages over the last few years. As much as you would want to believe that this is a one-time cost, lacking the proper experience can be a reason you lose your bees sooner than you would have anticipated. But, hopefully, this book will help you get it right from the beginning, so you won't have to worry about losing more money to pay for the steep price of bee packages.

- **Protective Gear:** As a beginner beekeeper, you need to make sure that you have all the protective gear you need to attend to your bees. You can find many retailers selling protective beekeeping gear at different price points. However, for a suit, with a netted veil and a pair of gloves, you should expect to pay between 75 USD and 170 USD. To make sure you are getting your money's worth, make sure you buy a suit of decent quality that can hold its shape after multiple washes.

In addition to the clothes, you will need a couple of tools to maneuver the hive frames and scrape off the honey. The frame lifter, honey brush, and other tools will probably add up to 150 USD. Another important item for beekeeping is the smoker, which you will need to use to calm your bees down if they start acting hostile. A good smoker with fuel should cost you about 50 USD or more.

While the costs listed above are only intended to give you a general idea about how much it's going to cost you to set up your own beehive, there are a few factors that you need to keep in mind if you want to save some money. First of all, you should consider a beekeeping starter kit. Many retailers sell everything a beginner would need at a significantly lower price. Alternatively, you can buy all the items separately from one retailer and ask for a discount. Talk with local beekeepers and follow their advice regarding when and where to buy your equipment; this should save you a lot of time and money.

2. Choosing a Location for Your Beehive

The second factor to consider before venturing into beekeeping is finding the right location for your beehive. Here are some tips to help you make this decision.

- **Somewhere Cool**: High temperatures can tire the bees as they have to exert a lot of effort to keep the inside of the hive cool. Placing your beehive in a shed can compromise your bees' honey production as they will be busy making sure that their queen is nice and warm. A great location for your hive would be under a tree to give it enough shade while having access to some sunlight at the same time.

- **Away From Dampness**: A damp hive encourages the growth of mold and fungi, which can pose a real danger to your bees and subject them to many diseases. You need to make sure that you place the hive somewhere dry and protect it from surrounding moisture that can seep inside. During the rainy season, it's a good idea to consider investing in a hive cover to ensure that raindrops will roll away and leave the hive dry.

- **Elevation**: Raising your beehive above the ground will stop the moisture from the soil from dampening it. It will also allow you to work comfortably on your hive and harvest the honey without straining your back.

- **Water Source**: Bees need to drink frequently, so it's a good idea to make their lives easier and place their hive somewhere close to a safe water source. It doesn't have to be a huge fountain or a big pool; a small pot of water can do the trick.

- **Somewhere Quiet**: Bees can easily get irritated by noise. If you want to keep your bees happy and relaxed, place the hive somewhere quiet. Find a secluded spot in your backyard away from the noise to set up your beehive.

- **Protected From the Wind**: Strong winds can knock down your beehive and aggravate your bees; or worse, kill them off. The perfect location for a beehive has to be shielded from the wind.

- **Accessible**: When choosing a location for your hive, give yourself enough space to walk and work comfortably. Don't cram it somewhere out of reach where you would struggle to move around.

- **Out of Predators' Reach**: You can check with your municipality to identify if there are any predators in the area like bears that can attack your beehive. If there are any, make sure you find a place for your hive that would be out of predators' reach.

- **Have Enough Spacing**: If you are planning on having more than one hive, you'll want to keep at least a distance of 2 to 5 feet away from the other hives. This will allow bees from different colonies to move freely without interrupting each other. The spacing is also important for you to have access to each hive, and as mentioned before, work comfortably without feeling cramped. One thing to note, though, is to avoid having the hive entrances facing one another.

If you think about it, finding the best location for your beehive placement is more about following logic than science. By applying the factors we discussed above, your job will be much easier, and you will be able to find a place where your bees will thrive and live a happy and healthy life.

3. Climate Considerations

The climate where you live will have a huge impact on your beehive and its well-being. Considering the climate is especially important if you are planning on turning your beekeeping into a business. Not only does the weather affect your bees' activity levels, but it also affects the quality and yield of honey that they produce. There are a number of climate-related preparations that you need to think about before you can start beekeeping.

• *In Cold Weather*

If you live in a cold climate, your job to beekeeping will be ten times harder. It's not uncommon for even experienced beekeepers to lose their colonies to frost. However, if you pay close attention and make the necessary adjustments, your bees will be able to survive.

1. Consider moving your beehive. Even though it is strongly recommended to refrain from moving the hive around, drastic times call for drastic measures. When the cold months are fast approaching, try to find a warmer place for your hive so that your bees can stay active and carry out their usual duties.

2. Shrink the hive. By reducing the number of your hive boxes, you will give the bees the chance to huddle for warmth. That said, the smaller structure will be more likely to withstand chilly drifts and protect the bees from freezing.

3. Keep the bees well-nourished. When it's cold outside, the bees will be discouraged to leave the hive in search of food. To make sure your bees stay well-fed throughout the winter and that they don't feed on the honey they produce, put some food in the hive. You can buy some fondant at the local store or simply prepare some at home to put inside the hive before the harsh winter months hit.

4. Frequently check on your bees. You don't need to do it every day, but at least a couple of days a week, you should check on your beehive to make sure that everything looks good. Add more food if it's running out and consider finding a warmer spot if the bees seem inactive or if you find a few dead ones.

• *In Warm Weather*

Although the warmer weather is relatively better for beekeeping, excessive heat can have long-lasting damaging effects on your beehive. While bees usually do a great job in keeping the inside of the hive comfortably cool, when the weather gets extremely hot, there are a few things that you can do to help them out.

1. Insulate the hive. Just like you would in the winter, it is equally important to insulate the hive from the inside to protect it from the rising temperature that can melt it.

2. Increase water supply more than you would during favorable weather. You should be extra keen on providing your bees with enough clean water to cool them off. Experienced beekeepers advise to place a bucket of water near the hive and fill it with some wine corks so that the bees can use them as standing steps to drink without getting their feet wet.

3. Use screened covers. Instead of using a regular cover on top of the hive, use a screened one to increase airflow and keep the inside ventilated.

4. Don't let the beehive overcrowd. You can remove one layer off of your 3-box Langstroth beehive to increase the airflow within and discourage the bees from clumping together and overheating.

5. Replace the metal roof. If your hive has a metal roof, it's a good idea to replace it with a wooden one since metal is a good heat conductor and immediately overheats.

4. Local Rules and Regulations

To become a beekeeper, you have to abide by some local laws and regulations to make sure that you are compliant and to avoid being subjected to legal implications. The rules mainly aim to ensure the public's safety as well as mitigate the potential harms caused by insects and pesticides. Regulations will vary according to where you live; however, there are some generic ones that you can benefit from knowing. Here are some rules to implement before practicing beekeeping.

1. Acquire a beekeeping license. Usually, you can get one by making an appointment with the Agriculture department. An inspector will visit your apiary to make sure that everything is in place and that you qualify for the licensure. It is important to note that your license will need to be renewed every year, so make sure that you don't miss your renewal date.

2. Hang the registration number where it is visible. Usually, the outside of the box of the hive is a good place to put your registration number on display where inspectors can see it.

3. Report any suspected diseases that affect your hive to your local authorities. This way, you can help other local beekeepers become aware of potential dangers.

4. Get the right permit to transport your hive. If you are planning on moving your hive across your state's borders, it is crucial to get the required permit beforehand.

5. Carry out accurate bookkeeping. Record all the expenses, sales, and profits related to your beekeeping activities. Not only will this help you keep track of your business, but you will be ready for any surprising audits as well.

6. Register for compensation benefits. In many states, when you register as a beekeeper, you can benefit from compensation in case of hive damage due to infections and endemic diseases.

7. Inform the authorities in case of hive disposal. If you decide to dispose of the hive or sell it, you have to notify the local authorities within one week from the disposal date.

8. Get your honey regularly tested. To comply with local rules, you have to be prepared to have your honey tested every once in a while to ensure its quality and that you can use phrases like 'organic' and 'chemical-free' when marketing your product.

Minding the parameters that impact your beekeeping activities is the first step in your exciting journey of becoming a legitimate beekeeper. As you have read throughout this chapter, it will take you a lot of time before you can cover the many details involved. However, this book will walk you step by step to simplify the process and make beekeeping easier than you would have imagined. In the next chapters, we will talk about the beekeeping tools that we mentioned in this chapter, but in more detail. You will also be provided with a DIY beehive guide and some insightful tips on how to care for and maintain your beehive.

Chapter 8

Beekeeping Tools

Beekeeping can be a complicated process, given that bees can be very dangerous if provoked. Bees should be kept in a safe place where they do not interfere with people or other animals around your home. Nonetheless, beekeeping is a wonderful experience as long as you know how to handle the insects. Harvesting organic honey is one of the most rewarding experiences of the process, seeing as honey is good for our health and is considered a natural remedy that can treat or prevent various conditions. Additionally, the wax obtained from honeycombs can be used in the production of different cosmetic products. There are also many other advantages of keeping bees, but to succeed in this venture, you need to have the right tools. This chapter outlines some beekeeping tools that you should have to make your endeavor achievable.

Bee Hives

As you're getting started, you need to consider getting the appropriate hives where the bees will live. There are different types of hives available on the market, so you should do some research and choose the type that will suit your needs. You need to decide if you will keep bees as a hobby or if you want to start a business venture. This will help you determine the right type of hives that you can get for your bee garden. When you have selected the right hive, you also need to choose the right frames where the bees will build their honeycombs. Again, there are different types of frames that will be discussed in more detail in the next chapter, so you should choose the right ones that will make it easier for the bees to start making honey.

Depending on your space, you should strategically position your beehives so that they are safe from natural predators like ants and beetles. It is a good idea to place your beehives on stands rather than directly on the ground. A hive stand elevates the hive from the ground to prevent it from weather vagaries like dampness caused by dew. This also helps keep your hives safe from weeds that can disrupt the movement of the bees. That said, you should always avoid getting beehives made of wood, seeing as termites can infect this material if you place the hive on the ground. A single hive stand can support more than two colonies. Hive stands can be made of wood, pallets, bricks, or concrete blocks, so you should choose something that suits your environment and homestead. You should also make sure that your beehives are secured by a perimeter fence to prevent strangers from accessing them.

Essential Oils

When you set your beehives in the right position, you can use essential oils to attract bees to the boxes. Essential oils can also be used to provide supplementary feeding to the bees during the time of the year when food is scarce. You can also use essential oils to drive out beetles from your hive since they can disrupt your bees and prevent them from making honey. The problem with beetles is that they quickly multiply and can force your bees out of the hive. You can purchase different essential oils such as lavender, lemon, and spearmint since they all serve the same purpose around your beehive.

Protective Clothing

If you are interested in keeping bees, the first thing that you should consider is protective clothing. A quality bee suit is a good investment since it is designed to offer you protection against the bees. There are different types of bee suits available on the market, and your ultimate choice is a matter of personal preference. You can get full bee suits that are designed to cover every part of your body from head to toe, and this helps to give you maximum protection. A full suit is often expensive, and it comes with different components, that include facial screens, to ensure that every part of your body is securely covered.

You can also get protective gear in individual pieces. Some people choose to wear just a veil and jacket combo. Different types of jackets are available in solid cotton blend styles, and others are ventilated. Additionally, you can also get pants separately from the jackets. Some protective gear consists of zippers, while other styles have elastics around their hems to offer a secure fit and tight protection against bees so that they do not find their way inside your suit. With the right gear, you can easily harvest your honey from your beehives without any issues. However, this is impossible if you do not have protective gear.

Jacket With A Veil

If you do not want to get a complete bee suit, you can consider a jacket with a veil that you can wear together with long pants. Bees have carbon dioxide receptors on their antenna, and these sensors allow them to detect human exhalations, which make them respond aggressively to protect the hive. Bees can also sense fear, and they can attack if you're feeling nervous as a beginner. However, some experienced beekeepers become so comfortable with bees such that they do not wear any protective gear at some point in their lives, after forming a bond with their colonies.

Bear in mind that bees do not wantonly attack, but they only do so when they feel that their hive is facing danger. It is a good idea to get the right protective gear to ensure that you are not stung by the bees when you handle them. Apart from getting the right bee suit, there are also other items that you can consider for your safety if you are interested in bee farming.

Shoes

It is crucial to ensure that your feet are properly covered before you dig through your beehive. You can achieve this by getting the right boots that are specifically designed for that purpose. When you are working on your beehives, it is important to wear flexible rubber boots, since they are designed to provide great protection to your feet. You should also ensure that your boots consist of quality material so that the bees cannot sting you through the material. As you are now aware, bees attack if they think you want to disturb their beehive so you should always be prepared.

Gloves

Gloves are also important since they offer protection to your hands. When you choose gloves, you should make sure that they consist of sturdy material and that they are long enough to cover your arm. You should also ensure that the ends consist of elastics for a tight fit to ensure that no bee will sneak its way to attack you. All the same, you should ensure that the gloves are designed to have proper ventilation to prevent your body from overheating while you are harvesting honey. While some people may choose not to use any protective clothing at all, you should know that stings are unavoidable, and they are part of keeping bees. It is always essential to ensure that the delicate parts of your body are fully protected.

There are different clothing options that you can consider, but you should know that bees often respond depending on how you approach their beehives. The bees will respond in peace if you are not aggressive, and you are at ease. The bees can also pick up on how apprehensive you are, and they can leave you to do your thing. The moment you try to approach the beehive, the bees become alert and are ready to attack. Therefore, when you choose your protective clothing, you should consider your personal needs, not what others suggest. Choosing to approach your colonies without any protective gear is something you can do eventually, after the bees have come to realize who their keeper is, and that your approach is not dangerous to their well-being. Bees can be dangerous, so always be on the safe side and get the right gear that can suit your needs.

Queen Catcher

The queen bee is responsible for laying eggs to ensure the survival and continuity of the bee colony. A queen is a necessary component of any beekeeping process. It is essential to identify the queen bee, and if you can't, you may need to re-queen. Alternatively, you may need to introduce a new queen to the hive to ensure its survival. To do this, you should learn how you can catch the queen and keep it separated from other bees in the hive if the need arises.

This is when a queen catcher comes in handy if you are a beekeeper. When you are going through your hives, you need to check if the hive has found a queen so that you can make an informed decision of whether to introduce a new one or breed one. However, luckily, the queen in the hive can stay as long as she wants since she is always protected. If you want to catch a swarm after introducing the queen to a hive, you can also use the same tool. You should know that bees may not come to your beehive naturally, and you may need to trap them and bring them to your hive.

In the same vein, when you are a beginner in keeping bees, it is important to have a queen marker that can help you identify the queen bee. You can use a bright marker to highlight her hindquarters so that you can easily identify her. This helps you to locate her if you want to transfer her to another hive so that you can grow your bee garden.

Feeders

At some point, you will need a feeder that can help you provide supplementary feeding to your bees. For instance, when plants have yet to bloom, you need to help your bees with a food supply so that they will not struggle to survive. You can do this by mixing equal parts of sugar and water that you dispense using a feeder. A feeder is not a sophisticated tool as you may think. You can use an open bucket with the sugar solution where the bees can easily access it.

It is important to buy your sugar in bulk so that your bees can multiply quickly to increase the number of your hives. If you have many hives, you should also increase the number of feeders so that your bees can easily access supplementary food. You need to place your feeders in strategic positions, away from your house, so that you do not interfere with the bees as they look for food. You should keep a close eye on the feeders so that they are free of ants, and you should also make sure that they are not accessible to other animals or pets that you keep at home.

A Smoker

A smoker is another invaluable tool that every beekeeper should have since it is designed to make aggressive bees docile. This tool is designed to confuse the bees so that they will think there is a veld fire nearby, which compels them to eat the honey quickly in preparation to move to another place. With full stomachs, bees become physically inactive, and they experience difficulties in tipping their abdomens in preparation to sting. This can help you to harvest honey without issues as the bees will be inactive.

The other advantage of using a smoker is that it acts as a mask to prevent the pheromone alarm system that is often raised by bees on guard. This will minimize the defensive reactions by other bees as they will not be able to communicate in a way that allows them to defend the beehive. When there is smoke in the beehive, the bees are temporarily disoriented, and they cannot sting because they are not able to send warnings throughout the hive. When the bees are disoriented, the beekeeper can easily go about their tasks of inspecting the hives, extracting the honey, as well as conducting frame removals without challenges such as bee stings.

There are different types of smokers, but the common one is made of stainless steel, and it also consists of a heat shield. A smoker also comes together with a solid chimney to improve burning. The smoke will gently float through the chimney into the hive, and you should wait for a moment before you start the extraction purpose to allow the smoke to spread across the entire hive. This will help reduce communication among the bees so that they become less aggressive.

Hive Tool

A hive tool is crucial since it is used for different purposes, particularly when carrying out inspections of your beehives. A hive tool is a flat, solid metal tool consisting of a tapered curved end and a sharp tapered end on the other side. These tools also consist of bright colors like yellow and red so that they are easy to locate in the bee yard or even at night.

The hive tools are mainly used for removing the frames that are heavy with propolis, also known as bee glue. This is a sticky substance that is made of tree resin. You can also use a hive tool to scrape the propolis away or open the beehive to harvest some honey. The tool can also be used to squash unwanted intruders like beetles as well.

Bee Brush

A bee brush can be more useful than you can imagine since it is used for various purposes. The brush consists of soft bristles, and it is used to remove the bees from the combs and other places that you do not want them to be, especially when you are harvesting honey. You can also use the brush to repair a broken comb, but you should use it gently since the bees can attempt to sting it. Do not apply force when you are using the brush since it can cause harm to the bees. Improper use of the brush can also anger the bees, which leads to a rise in pheromones and increased aggression. You should only use the brush when it is necessary to avoid causing problems that can affect your beekeeping endeavor.

Books About Beekeeping

Beekeeping is a dynamic experience that constantly changes depending on your environment. To succeed in this practice, it is essential to read many beekeeping books so that you can gain insight into the necessary measures that you can take to achieve your desired goals in more detail. Your honeybees can be affected by different environmental factors that you should know, and you can gain more knowledge about beekeeping from different books. Another thing that you should know is that every honeybee colony is different, so you should make appropriate management decisions based on the colony that you have.

When you decide to keep bees, you should be flexible and be able to study why bees behave in a certain way. You should also know how certain actions can impact the wellbeing of certain bees. More importantly, you need to exercise patience when you are beekeeping. There are many steps involved, and progress does not happen overnight. You need to understand the best beekeeping practices and get the right tools that can help you achieve your objectives. Additionally, you can also consult experts in the field of beekeeping so that you can do the right thing to avoid frustrations that can put you off track.

Extracting Equipment

Honey extractors often work the same way, whether they are used for small scale or commercial purposes. You can get a manual or electric honey extractor. They consist of stainless steel bodies that are cylindrical, and they have many baskets that are designed to hold the frames of honey. The honey is pulled from the frames using centrifugal force, and this applies when you are using an electric or manual extractor. The honey drips down the inside wall of the extractor, where it moves to a spigot placed at the bottom. The honey will then run through a food-grade strainer, where it should be allowed to rest for about 24 hours in the bottling buckets before it is bottled.

If you are a serious beekeeper and you intend to harvest honey several times every year, you should invest in extracting and bottling equipment. You can enjoy the spontaneous extraction of honey during any time of the year as long as you are certain that it is ready for harvesting. Alternatively, you can rent honey extractors from your local store, since this can be a cheaper option. What is important is to ensure that you properly extract your honey.

Besides getting the right tools and equipment for beekeeping, it is essential to operate with a budget that can help you get the items to suit your needs. Some bee tools come in the form of sets, and their prices significantly vary depending on factors such as quality and size. When choosing the tools, you should also understand their purpose and how to use them so that you do not waste your money on items that you may never use. The number of colonies that you intend to keep can also help you get the right equipment.

Just like any other hobby, if you are interested in beekeeping, you should invest in the right tools and equipment that can make your endeavor easier. Beekeeping has evolved over the years, so you need to do some research that can help you understand different things that are involved in this venture. While there are different beekeeping tools available on the market, it is important to choose the equipment that can cater to your needs. You need to know that items that can work for other people may give you different results. So, you should stick to what you think can help you achieve your goals with fewer hassles. With the right tools, you can make beekeeping an exciting hobby or a lucrative business venture. All the same, you should always remember that bees can be very dangerous, so make sure that your bee garden is located in a safe position that is safe from human and animal interference.

Chapter 9

DIY Beehives

A s you have read so far in this book, we are providing you with all the tools that you need to build your own backyard homestead. After having discussed the relevance of beekeeping and how it will have such positive impacts on your garden's yields, it's time to talk about DIY beehives. By the end of this chapter, you will be able to identify everything you need to get started on your backyard beehive. Whether you are thinking about selling your honey harvest or you simply want to self-suffice, it's an interesting project that doesn't require a lot of time or money. Continue on reading the below to find out everything you need to know about DIY beehives.

1. DIY Beehive

Building your own beehive means that there isn't just one way of getting it done, you can use different tools and materials as long as the hive ends up adequate for your bees. Different kinds of beehives include:

• *Langstroth Beehive*

As we mentioned earlier in chapter 5, the Langstroth beehive is one of the most popular favored by beginners and seasoned beekeepers alike for a number of reasons.

1. It has a stackable structure, which means that you can add and remove layers as you see fit.

2. It is widely common, so it will be very easy for you to find ready-made accessories and fittings for your hive.

3. It requires little to no maintenance.

4. It will save you space. This is an essential feature if you have limited space in your backyard.

However, in order to make the right decision, you also need to understand the drawbacks of the Langstroth beehive. It is usually heavy, so you might not be able to change its location once you set it up. In addition, a typical Langstroth beehive structure doesn't have any openings to allow you to view its insides.

To build a Langstroth beehive, you will need to have sturdy wood logs and some basic tools like a table saw, drill, and tin snips. Here is a general overview of what a beehive comprises.

1. The Langstroth beehive consists of a bottom drawer, which is the only place your bees can enter and exit.

2. An entrance reducer is more or less an accessory; however, you will need it to control the temperature inside the hive, especially during winter, when you'd want to protect your bees from the cold weather. However, during summer, when it's honey harvesting season, you won't need the entrance reducer, and it's better to remove it completely to allow for more honey production.

3. The hive body is where your bees live. You can build two hive bodies, one for the bees and the other one for their food. If you live in a relatively colder climate, experts advise using only one hive body to keep the bees warm.

4. There's also a queen excluder in every beehive since the queen bee doesn't partake in the honey production phase. You will need the queen excluder to separate between the hive body and the honey super, which is placed on top. The queen excluder will also avoid having the queen bee lay eggs in your honey yield.

5. The honey super is identical in shape to the hive bodies, but it's usually shallower. This is where the honey that you will get from your bees is produced. As a beginner beekeeper, you

won't need more than one honey super box. However, as you add more bees to your colony, you can add more honey super drawers to accommodate the corresponding increase in honey production.

6. In every beehive, you'll find frames. Bees use frames to build the honeycomb. You need to make sure that the frames are equipped with a proper foundation sheet made from either beeswax or plastic.

7. The inner cover of the hive is a simple wooden tray with a hole in the middle. This allows air to pass in and out of the hive.

8. The outermost layer used to protect your bees from the outside world is the top cover. For extra protection against changing weather conditions, you can attach an insulating material sheet made of aluminum flashing to the top of your outer cover.

This is a simple project that you can get done over a weekend. Search online to find a step-by-step guide to walk you through the process if you don't consider yourself a great handyman. Building your own Langstroth hive will save you a lot of money and will allow you to customize your beehive to fit your personal needs.

• *Top Bar Beehive*

The top bar beehive is another variation of beehives that you can build by yourself in your backyard. This beehive is one of the oldest styles that beekeepers have been using for ages. The Top Bar beehive looks like a small dining table with wooden bars laid on top of the hive cavity, and it usually has a rectangular or tub-shaped body. Bees use the guided bars to build their combs and let them hang straight. Top Bar hives are easy to handle and are less complicated than Langstroth hives. The main benefits of Top Bar beehives include the following:

1. The simple structure of this kind of beehive means that it's extremely cheap to build.

2. It is mobile thanks to its lightweight and minimalist structure.

3. It requires minimum maneuvering to maintain, which means that your bees will have minimum disruptions.

4. The wood bar combs are easy to move and allow for easy inspection.

5. It provides better observation than the Langstroth due to its flat horizontal setup.

6. It reduces the possibility of getting stung by the bees, so it's highly recommended for beginner beekeepers like yourself.

However, on the downside, Top Bar hives require frequent maintenance work and are known to produce less honey than other hive types. Furthermore, if you want to grow your bee colony, you will require a huge space to accommodate it if you are using a Top Bar hive.

To DIY your Top Bar beehive, you'll need to do the following:

1. You can use an old plastic barrel cut in half to pose as the hive body. Make sure that you clean the barrel well and rub its insides with beeswax to make it more appealing to your bees.

2. Using sturdy wood lumber like cedar or plywood, use your saw to create a frame to cover all four sides of your barrel, then attach it using some screws and heavy-duty wood glue that is temperature-resistant.

3. According to your own height, build the wooden legs to carry the barrel at a suitable height that allows you to work comfortably without straining your back.

4. Screw the wooden legs to the barrel and make sure that they are securely attached.

5. Measure the length of your barrel and cut wooden bars that will be used by the bees to build their combs. Make sure that you add a guide so that the honeycombs come out straight and don't collapse. You can attach a beeswax-coated twine to each one of the bars.

6. Build the roof of the hive using a tin sheet, but make sure you leave about a quarter of an inch on each side to ensure roof properly closes off the barrel.

7. Secure any extra bits of the tin roof by screwing them to the sides of the wooden frame circumference the barrel.

8. To make sure the roof stays put and doesn't fly away; you can use a tie wire to tie it to the hive body.

Top Bar hives are simple and easy to build. You can manage to get yours done in a matter of hours.

• *Warre Beehive*

A Warre beehive looks a lot like a Langstroth hive on the outside. However, Warre hives have some distinctive features that make them unique. In general, Warre hives consist of 3 main layers tiered vertically on top of each other. Each of these layers, the base, the boxes, and the roof serve a purpose. Warre hives are great for the following reasons:

6. They provide the most natural environment for the bees.

7. They are the easiest to maintain and are considered to be lightweight compared to the similarly shaped Langstroth hive.

8. Warre hives allow for minimum disturbance for the bees.

9. The vertical orientation of the Warre hives makes them space-efficient.

Building a Warre hive will cost you more than a Top Bar or a Langstroth hive would. Furthermore, they are not as lightweight as the Top Bar, which makes it hard for you to move them around if you need to.

With intermediate carpentry skills, you can build a Warre hive by following the below steps:

6. Start with building the Hive Body. Using quality wooden logs, you can build the hive body with the dimensions 300X300X210 millimeters to create the space where your bees will spend most of their time.

7. You can then proceed to build the Topp Bars. You can cut 8 24X315 millimeter bars of wood to create the top bars on which the bees will build their combs.

8. The floorboard would be your next step. It basically supports the entire hive, so you have to make sure that you do a great job with it. It's strongly recommended to build wooden legs from treated wood to carry the Warre hive instead of placing it directly on the ground. This will keep your hive safe and out of the reach of curious animals like rodents. The elevation will also protect both the floorboard and the ground underneath. To build the floorboard, you can cut a 338X338 millimeter wood tray with a minimum thickness of 15 millimeters. You can add a landing board underneath to make it easier for the bees to enter the hive from the outside.

9. The quilt box, on the other hand, is the layer that lies right underneath the roof. You can use the same measurements that you have used for the hive body to build the quilt box. However, you don't need to make it as thick since it will only carry the weight of the roof. Fill your quilt box with wood

shavings to control the moisture inside the beehive and to help insulate it against external weather conditions.

10. You can use either wood or metal to build your gabled rooftop to let the raindrops and snow roll off the hive. It's important to create a ventilation opening in the roof to keep the hive cool during the summer. Typically, the roof should be 120 millimeters deep. If you find it easier, you can build the quilt box and the roof as one single structure.

The best thing about Warre beehives is that you don't need to check on them constantly. They require minimum intervention on your side and provide your bees with the best environment to do their own thing.

Simpler DIY Beehives

If the above models seem a little intimidating, don't worry, we have other alternatives for you to choose from. Here are some ideas for simpler DIY beehives that you can "build" from upcycled items that you can easily find at home or buy from a thrift shop.

• *Old Computer Case*

The ready-made structure means that you will need minimal carpentry to build a basic cover for your beehive. Due to the limited size, this is a starter beehive that will only accommodate a small bee colony. However, it's a great place to start experimenting with beekeeping and hone your skills.

• *Mason Jar Beehive*

Mason jars are staples in DIY projects. It might come as a surprise, but you can actually use these versatile jars to build your own quirky beehive. The best thing about this alternative is that the honeycomb and the honey will already be inside the jars, saving you the hassle of harvesting the more traditional beehive structures.

• *Nucleus Beehive*

Think of the nucleus beehive as the compact version of the standard Langstroth or Warre beehives. You can use it to start with a few frames until you get the hang of beekeeping, and you can then move onto more legitimate versions.

• *Log Beehive*

Using the log from a collapsing tree in your backyard, you can create simple hollows for your bees to nest. This is the cheapest DIY beehive that you can possibly build.

• *Old Tires Beehive*

Since this book is meant to guide you in becoming a professional homesteader, you are encouraged to explore new ways to reuse and upcycle items that you already have. Using a couple of old tires, you can build a home for your bees to thrive and produce the best organic honey at minimum costs.

2. Adding the Bees

Now that you have many beehive options to choose from, it's time to add your honey bees. You can either buy or catch the bees.

• *Buying Bees*

Usually, you have a couple of options for buying bees.:

1. Package bees: This is usually around 2 to 5 pounds of bees, including the workers and one queen bee sealed separately. The package also comes with a syrup-filled feeder to feed your bees. You can get package bees online or from a local supplier who will give you all the information you need about how to install the bees in their hive and how best to introduce the queen.

2. Nucleus hive: Nucleus, or nuc hives as they are commonly known, are usually made up of 5 frames, including the honeycomb, the bees, the baby bees (brood), and one queen. Although buying a nucleus hive allows you to get right into harvesting your honey, it is considered to be a riskier approach as you can never be sure about the health and safety of the donor's hive.

• *Catching the Bees*

If you are looking to save money and your state laws allow catching wild bees, you can harvest a swarm of bees from the wild. During spring, you will find clusters of bees traveling together in search of a new space to accommodate their growing colony. Although collecting bees while swarming is relatively easy, as the bees tend to be in a calm mood, you need to make sure that you are wearing the right clothes in case you get stung. In addition, having a smoker at hand might come in handy if the bees start to get nervous. Experts recommend that you consult your local beekeeping association to help you find the best ways to ensure your swarms' health and that the queen is alive and healthy.

3) Beekeeping

Now that you have your beehive filled with bees, the real work starts. You can register to attend online beekeeping classes or join the local beekeeping association to make sure that you have enough knowledge about keeping your bees healthy and happy. Connect with experienced beekeepers and ask if you can shadow them to watch how they manage their hives and learn about the small tips and tricks. After some time, you will be able to master beekeeping and understand everything you need to do for a successful endeavor.

Besides the numerous benefits that we mentioned in previous chapters about beekeeping, it can also be an interesting hobby. Caring for your bees will teach you a lot about nature and appreciate the simple things in life. Now that we are approaching the end of this book, the following chapter will provide you with important information about the best practices in beekeeping. You will also get to learn about the tools you need to maintain your beehive so that it can thrive and grow further every day.

Chapter 10

Best Practices and Maintenance

Every beekeeper cares about the wellbeing and balance of their beehives and strives to maintain the colonies to the highest of standards. In order to make sure that the colony is well-maintained, certain practices need to be followed so that the honey bees are living in the best possible conditions. Unfortunately, in today's world, it has become more and more challenging to maintain the wellbeing of the hive and ensure that the bees are remaining productive and healthy, seeing as pollution along with other environmental and human factors can deter that. However, it is not impossible to offer good care for your honey bees and maintain your hives to quality standards that would be not only profitable to you as a keeper, but also beneficial to the planet as a whole. Here are some of the best practices and maintenance tips that can get you started with your hive of honey bees.

Nutrition for Bees

The first thing that should be on any beekeeper's list of essential maintenance and best practices for bees is knowing how to keep the hive alive and in good health. Bees are just like any other living being; they need good nutrition in order to survive and be productive. In fact, bees are among the most productive creatures found on earth, and they need sufficient nutrition to keep them going strong and even increase that productivity in some cases, depending on what they are being kept for. A successful beekeeper should always keep an eye on what kind of nutrition their bees are getting and how the hive is responding to that kind of nutrition. There are several ways to keep the bees in good health when it comes to their nutrition, and they each depend on different environmental and physical factors.

• *Natural Forage*

Bees have been surviving on their own, and still are, without any human intervention in their keeping. They are very smart creatures that know how to take care of themselves and feed themselves accordingly with balanced nutrition. That is why in many cases of beekeeping, keepers can simply rely on natural forge as a form of nutrition for the bees where they can move around and get the food they need from the crops surrounding them. The beekeeper's job, in this case, would be to simply place the beehive in a suitable location where there is a healthy diversity of natural pollen in floral collections that would allow the bees to feed themselves safely and depend on themselves for nutrition. The beekeeper should also avoid overcrowding one location of floral crops with too many bees and a large number of hives. They should instead divide the hives, if there is more than one, into separate groups in different suitable locations so that the bees can get the sufficient nutrition they need without competing for the pollen.

• *Supplemental Feeding*

In certain locations, the weather conditions can either be very extreme or quite favorable, and the beehive might find it challenging to get the right kind of nutrition in traditional or natural ways. In such circumstances, the beekeeper should work on providing supplemental feeding for the colony, where they provide protein pollen patties for the bees to feed on for the almond season if they are honey bees. That said, protein pollen patties are essential for beekeepers looking to build strong colonies in non-suitable weather conditions, where they can massively enhance the hive's survival and performance rates. Protein patties are extremely nutritious and easy to absorb for the bees as they do not pose any artificial health risks on the colony, but rather help improve the bees' overall well-being.

• *Hydration*

Just like any other living being on planet Earth, bees need water to survive. Beekeepers need to provide an abundance of water sources for bees to find and to remain hydrated, especially during drought seasons, where many colonies can face higher risks of death by dehydration more than anything else. Even if there is a natural water source in the surrounding area where the beehive is located, it is the beekeeper's job to ensure that that source of water is clean and safe for the bees to consume. Seeing as many natural water sources can be infected with insecticides or harmful fertilizers that can be deadly for the bees. The beekeeper can build their own artificial water sources that would easily be accessible for the bees so that they can carry on seeking pollen while staying well-hydrated.

Pest Control

A beekeeper's worst nightmare is finding pests in the hive. The hive is the bees' home, and just like any home, it needs regular maintenance and a high standard of cleanliness to ensure that the bees are safe and are in good health. To avoid pests sneaking into the colony and making their way into the hive, numerous actions should be taken by the keeper. Different types of pests and viruses are commonly known to affect beehives, and some of them can be extremely dangerous and even deadly if they're not caught early on and eliminated effectively. Certain kinds of toxins and the Varroa virus are among the most common risks any beehive can face.

• *Checking Methods*

Controlling pests in the hive starts with doing regular check-ups on the colony. Toxins can be easily spotted using professional equipment that can detect any poisonous air or lingering insecticides. Checking for Varroa, however, can be a little more challenging, yet equally essential. You can check for the presence of Varroa using sticky boards to accurately determine the mite count and work out how they can be dealt with accordingly at a later point. The beekeeper might also need to use an alcohol wash along with sugar or an ether roll in order to be certain where the Varroa is located in the hive and how it can be eliminated.

• *Eliminating Methods*

Once you check for pests and find them in the beehive, the next logical step would be eliminating them and working on ways that ensure they can never make their way back. In the case of Varroa, using powdered sugar dusting and certain soft chemicals is likely to do the trick and eliminate the virus entirely. Varroa mites might make a comeback during the summer season if the treatments are not done on a regular basis, so you need to ensure that the mites are gone before they start doubling in numbers. On the other hand, when it comes to other kinds of toxins, being aware of the kind of toxin exposure near your colony is key to eliminating any problem before they develop further. Renewing the beeswax every now and then can also ensure that any toxins found the hive is removed every year.

Maintaining the Equipment

Beekeepers use a variety of tools and equipment when it comes to maintaining the hives and caring for the bees. In order to be able to care for the bees and maintain the hives to the highest of standards, it is essential to take on the practice of maintaining the equipment itself. This can be done by regularly checking for rotten frames or loose ends in the hive and quickly fixing them before they collapse with the colony. It is also important to make sure that your beekeeping attire is in good shape so that you do not risk your own health in the process of beekeeping as well as that of the bees. It may be wise to do most of the maintenance work on the hives during the winter months as it makes things easier for the keeper as well as the bees if any replacements need to be made.

Hive Security

Having a beehive on your farm can be extremely valuable. Honey bees, in particular, are well sought after and thieves will have their eyes on them. That is why making sure you practice the ultimate standards of hive security is massively essential, especially if you have a large number of hives on your property. As the pollinating value of crops on your property increases, the risk of hive theft increases as a result. To ensure your beehives are safe and kept safe from the risk of theft, make sure you keep all your hives IDed or numbered in a way that makes it easy to identify them. You should also check them on a regular basis and install the needed surveillance systems that notify you of any breach. It is also essential to practice discretion when it comes to showing different individuals around your property, or the yard where the beehives are kept.

Monitoring the Colony

A successful beekeeper should always know all the ins and outs of the colony and every small detail concerning the hives. Monitoring the colony means knowing the needs of every group of bees you have. These check-ups are something that needs to be done regularly and actively to ensure that the hive does not lose its productivity and strength under the beekeeper's nose. If one colony shows signs of weakness, it is essential to keep it separated from other strong ones and focusing on bringing that colony back to its full strength, while working on analyzing and understanding the causes of the problem to prevent it from happening again.

Wisely Managing Stocks

Maintaining beehives is no easy task, and it involves a lot of tools, equipment, as well as stocks of everything needed for a high-quality standard of colonies. Stocks, in this case, are not only items used in caring for the bees, but they also include stocks of bee colonies, so choosing the right types and maintaining them effectively are crucial steps to properly managing your hives. Some bee colonies can be extremely aggressive rather than productive. Those colony stocks are not the ones you need on your property, as there would not be any upside to keeping them. It is rather wise to keep your existing colonies productive by annually re-queening them to ensure that productivity does not decrease and that pollination never stops.

Best Maintenance Practices for Beekeeping as a Business

Some people keep bees as a kind of hobby or for personal purposes. However, if you are keeping bees to start or work on a profitable business, your maintenance scale and the kind of practices you or any employees do need to be professional and efficient. This will ensure the success of that business, and having contracts that cover every small bit of detail regarding any deals done concerning the bee colonies is an essential step that should be taken by any business beekeeper. It ensures the safety of all parties involved, as well as security for the beehives.

• *Efficiency*

Any business requires a high standard of efficiency to be practiced, maintained, and to ensure its success. The same thing goes for the business of beekeeping. It is essential for professional beekeepers to be fully aware of all the details concerning their colonies and to work on strong records that can attract growers and other interested parties to their beehives. Using quality equipment and following through with high standards of care is also essential in maintaining efficiency for the business as well as keeping the bee colonies productive.

• *Continuous Learning*

Beekeeping is an educational experience that beekeepers never stop learning from. There are new developments in the field of beekeeping maintenance and smart practices being created constantly that can help business owners lead a more successful beekeeping business. There are numerous beekeeping journals and scientific articles constantly being released with all the essential developments that beekeepers could benefit from, so make sure that you keep yourself updated with beekeeping news and methodologies. If you are a beekeeper, it might also be worthwhile to join beekeeping groups and organizations to network with and learn from other professionals on the same boat as you.

• *Creativity and Inclusivity*

Running a successful beekeeping business requires being creative, as today's market is full of innovative ideas and technological advancements that make it hard for anyone to stand out in the crowd. To get that extra edge of creativity as a beekeeper, it is important to give back to your community by including new talents and working with them on incorporating new advancements to your traditional beekeeping strategies. Including new talent means giving back to the industry and staying creative. Beekeepers can also contribute to new research on bees and colonies and allow scientists to monitor their hives as well as come up with new research that can benefit everyone, including yourself and your environment.

Bees and Agricultural Practices

Growing crops and beekeeping are two things that go hand-in-hand in a very harmonized manner. Maintaining beehives alongside agricultural crops depends heavily on the placement of the colonies in relation to that of the crops so that the pollination process can be done easily with smooth accessibility and convenience. It is also essential for beekeepers to keep an eye on the kinds of chemicals used in any nearby agricultural areas as in the majority of cases, such chemicals can be extremely toxic for the colonies. Ensuring that the process of caring for the crops safely while simultaneously watching over the bees and where they go is essential for the wellbeing of the bees and the success of the business.

Pollination Essentials

The pollination process is usually an agreement between a grower and a beekeeper. It is not too common to find one business owner who is both. So when it comes to pollination, both parties profit from the process. This is why it is an essential practice to sign fair contracts for both parties, where all the details are clearly set out for everyone to agree on and benefit from. It is also important that the beekeeper is around and keeping an eye on the colonies during the pollination period to see the bees in action and monitor the process accordingly.

Safety with Bees

Bees are very friendly beings; however, they require very careful practices when it comes to caring for them and maintaining their hives to ensure the beekeeper is safe and away from any risks. A professional beekeeper should always wear suitable attire when approaching the hive and practice impeccable hygiene so as not to infect the colony with any bacteria or other toxins. All equipment used with bees needs to be regularly cleaned and disinfected with soft chemicals that would not harm the bees in any way to help the beekeepers carry on their work in a safe environment.

Beekeeping is a practice as old as time. Bees are very productive living beings that maintain the balance on our planet, and so it only makes sense that beekeepers need to strive and work hard to maintain a high-quality life for the wellbeing of their bee colonies. Beekeeping is an educational journey, so as a beekeeper, it is important to keep on learning and researching as long as you are trying to care for bees and maintaining their hives to ensure you get the most rewarding experience in beekeeping.

Conclusion

A homestead is a pretty big investment in your future, revolving around the sustainability, experience, and labor that you put in. Various homesteading approaches vary in difficulty and the required level of experience. This was taken into consideration as this book was written with complete beginners and non-experienced enthusiasts in mind.

Your first practical contact with the homesteading experience will be through choosing the right equipment that you're going to need. The type of gear that is required is mainly dependent on the type of homesteading that you're going to do. Whether you'll be using an ax to chop wood for a fireplace or a tractor to farm many acres, you'll want to ensure that you invest enough time in understanding the need for these tools in your specific homesteading situation.

Since the book's focus is mainly gardening and beekeeping, it's important to ensure that you have gloves, hand trowels, and shovels for gardening. Hive tools, bee brushes, magnifying glasses, drones, queen marking kits, and many others. This is why it's important to carefully read the chapter on the tools and equipment needed for homesteading, as it can easily get confusing when you're buying equipment for the first time without a compass to guide you. You also don't want to buy a tool once the occasion for its use arises; it's better to be prepared because rushing out to buy equipment is usually impractical.

That said, setting up a backyard garden requires a careful evaluation of your needs, while also taking into account the need to scale it further or diversify it later on. You don't want to waste your money on setting up the wrong kind of backyard homestead for your needs. This is why property size, type, and conditions are the most important factors that define any homesteading project. A backyard that's more suitable for gardening shouldn't be utilized as land to raise livestock on. Rural areas that don't have stable sources of electricity may need a generator or solar panels to stay powered more efficiently. You'll notice that you'll be making your decisions based on the final image of the homestead you have in mind, so make sure it's elaborate enough to see it through.

Of course, having little experience with flowers can make the process of choosing them for your backyard garden a more difficult process than it actually is. You'll want to make sure that you include pollinator-friendly plants, since you may want to attract bees for your backyard beekeeping. Pollinator-friendly plants play an important role in the ecosystem of any garden, making it your top priority when you're choosing the right types for your garden. Flowers are often categorized into three different seasonal categories. They are annuals, which can bloom for the whole season and some of which are even self-seeding; biennials, which take an extra year to bloom;

perennials, which last the longest but are the most expensive as they tend to die off quickly.

Keeping track of the logistics and dimensions of your future garden is essential as well. You are not just planting flowers for decorative purposes; you're looking to provide some sustenance by utilizing the products of the plants you're going to plant in the garden. This is why this book focuses on climate conditions when it comes to choosing the crops that your garden is best suited for. This is one of the most enjoyable aspects that you'll experience in homesteading. Designing a garden can make you indulge in both western and eastern gardening traditions. Finding creative ways to define your garden's borders visually can be a challenging but fun experience, nonetheless.

The garden you design is going to have a microclimate that can be taken advantage of if you choose the right crops for it. You can have a look at your neighbors' gardens to see what kinds of plants look the healthiest and to know the native species that can work in a similar climate. You don't want to plant a tree of the wrong size in the wrong location!

Something as simple as the tomatoes that take a few days to be transported from the farms to the grocery stores and finally to your doorstep will begin to taste quite bland when you compare them to the tomatoes that you're harvesting from your backyard. You'll be consuming a lot more vegetables in general when you're the one planting them. This is because you'll get to taste these vegetables in the best possible state, and you'll also enjoy eating something that you've worked hard for.

On the other hand, the vitamin content of the vegetation that you harvest and eat from your backyard is going to be much higher than that which has been treated with toxic pesticides and artificial fertilizers. Even the physical exercise you'll be doing to plant flowers, vegetables, or trees is going to keep your body in shape for a long time to come, as long as you maintain it. A lot of people treat backyard gardening as a stress-relieving activity that allows them to bask in the sun and help them feel rejuvenated.

With that being said, when you're in control of the crops and the land, it's easy to move crops to another area if it's not working out as you intended it to. You don't need a crew of experienced people to move your plants from one area to another when you have a backyard homestead. This is also quite beneficial for crops that are vulnerable to pests, as the small area won't make it inclined to get attacked by insects in swarms like what happens in farms. This also means that you'll be able to maintain it easily without a lot of additional costs.

And of course, those who are familiar with raw honey know how expensive it can be. It's known for its helpful enzymes that are usually lost when they are packaged conventionally to be sold in grocery stores. The organic honey that you harvest yourself from your very own homestead, on the other hand, is quite safe to consume if it's harvested right, with the help of the famous antibacterial properties that protect it from germs and bacteria when it's stored.

Aside from the delicious raw honey, you'll have access to; beekeeping was chosen for this book for a myriad of its benefits. Nothing can make your homestead as financially sustainable as beekeeping could thank its great financial value. However, you shouldn't sell your bees short and mistakenly think that honey is the only product that you can get out of them. Wax, propolis, and royal wax are amongst the most expensive products that bees produce easily. Some beekeeping homesteaders decide to become queen breeders to supply interested local beekeepers with bees. On the other hand, bee venom is one area that a lot of experts see the potential in because of how it uses in modern medicine, making venom production a viable and profitable specialization as well.

One element that makes beekeeping and gardening one of the best combinations for both beginner and expert backyard homesteaders is how they synergize and complement each other. A big portion of all crops produced in the world relies entirely on pollinators, which are mainly bees. The ecological effect of protecting bees and providing them with a safe environment that allows them to breed will make your business venture beneficial for yourself as well as the environment. As a beekeeper, you'll learn how a whole species of plants or insects can drastically affect the environment.

If you're thinking about expanding your backyard beekeeping, you might as well take into consideration the tax benefits that you can actually get from this venture. Agricultural tax benefits are provided to those who have more than 50 hives, which is a bit of a bigger operation than what beekeeping hobbyists handle, but it's still plausible. As a hobby, it's still a very rewarding experience that allows you to enjoy raising beneficial creatures without the hassle and expenses required to raise livestock.

In addition, as a new homesteader, there are a few things that you should always keep in mind as you initiate any project. As a rule of thumb, you should never take on a project because it seems simple; a homesteader's life can be simple, but it's not exactly easy. Expect to face a lot of hardships as you try to balance your new life with your career and personal life, especially if you're a total beginner. It's also better to have someone more experienced by your side to help you learn the ropes quickly, whether it's a neighbor or a friend.

There is nothing that can stop you from expanding a simple backyard homestead into a full-on homestead on many acres, but it's important to make sure that you perceive the learning curve properly. Don't take on complicated or expensive projects at the early stages of your journey; first, ensure that you have received enough training to succeed and to avoid wasting a lot of money and labor. The detailed guides for gardening and beekeeping in this book should make your endeavor a successful one while also providing you with creative input.

References

https://homesteading.com/best-homesteading-tools/

https://www.survivalsullivan.com/homesteading-tools-you-need/

https://www.thespruce.com/compost-bins-and-how-they-work-2131027

https://www.motherearthnews.com/homesteading-and-livestock/top-20-homesteading-tools-zmaz01amzsel

http://www.imperfectlyhappy.com/5-easy-steps-into-backyard-homesteading-2/

https://www.tenthacrefarm.com/start-a-homestead/

https://homesteadandchill.com/how-to-start-a-homestead/

https://grocycle.com/how-to-start-a-homestead/

https://morningchores.com/irrigation-system/

https://www.homestead.org/frugality-finance/small-scale-homesteading/

https://homesteadandchill.com/easy-annual-companion-flowers/

https://homesteadandchill.com/top-23-plants-for-pollinators/

https://www.wildhomesteading.com/perennial-biennial-annual-plants/

https://homesteading.com/types-of-flowers-to-plant-summer-flowers/

https://www.goodhousekeeping.com/home/gardening/a32638/sunflower-fun-facts/

https://morningchores.com/farm-layout/

https://www.almanac.com/content/garden-plans-homesteads-and-small-farms

https://www.sugarmaplefarmhouse.com/planning-a-homestead-garden/

https://thetinylife.com/basic-tips-for-homestead-gardens/

https://www.thehomesteadgarden.com/how-to-plan-your-garden/

https://www.thespruce.com/designing-vegetable-gardens-1403407

https://www.thespruce.com/prepare-your-soil-3016982

-https://farmhomestead.com/gardening-methods/

-https://learn.eartheasy.com/guides/raised-garden-beds/

-https://www.justdabblingalong.com/how-to-plant-homestead-vegetable-garden-easy-steps/

-https://savvygardening.com/cold-frame-gardening/

-https://homesteadandchill.com/easy-annual-companion-flowers/

-https://homesteading.com/gardening-tips-and-tricks/

-https://dengarden.com/gardening/How-to-Take-Care-of-your-Flower-Plants

https://www.theprairiehomestead.com/2014/05/get-started-honeybees.html

- https://commonsensehome.com/homestead-bees/

-https://www.almanac.com/news/beekeeping/beekeeping-101-why-raise-honeybees

-https://www.almanac.com/news/beekeeping/beekeeping-101-supplies-clothing-and-equipment

-https://www.kelleybees.com/blog/kelley-beekeeping/thinking-keeping-bees-part-1-costs-time-intangibles/

-https://beebuilt.com/pages/langstroth-hives

-https://backyardbeekeeping101.com/beekeeping-cost/

-https://www.keepingbackyardbees.com/8-proper-beehive-placement-tips/

-https://beekeepclub.com/tips-for-beekeeping-in-cold-climates/

-https://www.keepingbackyardbees.com/protect-your-bees-in-hot-weather-zbwz1807zsau/

-https://baynature.org/article/helping-bees-beat-heat/

-https://www.beekeepers.asn.au/news/2017/10/19/new-rules-of-beekeeping-made-simple

https://www.housebeautiful.com/uk/garden/a585/top-tips-garden-maintenance/

https://www.lovethegarden.com/uk-en/article/7-lawn-care-tips

https://morningchores.com/vegetable-garden-care/

https://www.thespruce.com/how-and-when-to-prune-plants-1403009

https://www.finegardening.com/article/10-ways-to-keep-your-garden-healthy